SHADOW ON THE SUN

June 2017

Book 1
of
The Hanford Trilogy

by

R. Julian Cox

Northern Lights Publishing Ltd

Shadow on the Sun

First published in Great Britain in December 2012
by
Northern Lights Publishing Ltd.
E mail address:
info@northernlightspublishing.co.uk

Copyright ©2012 by R. Julian Cox

Fourth reprint

The Author asserts the moral right to be identified as the author of this work

ISBN 978-0-9573226-0-8

All rights reserved. No part of this book may be used or reproduced in any manner whatsoever without written permission, except in the case of brief quotations embodied in critical articles and reviews.

For further information address
Northern Lights Publishing Ltd.

Copy editor: Mark my Words (natilebs@gmail.com)
Story editor: Winskilleditorial
(www.winskilleditorial@btinternet.com)
Cover artwork by Kennzi
Set in 11pt Adobe Caslon
Production: Electric Book

Printed and bound in Great Britain by
Lightning Source (UK) Ltd., 6 Precedent Drive,
Milton Keynes,
Buckinghamshire
MK13 8PR

Available in hardback, paperback editions and ebook formats for use with such e-readers as Amazon's 'Kindle' or Barnes and Noble 'Nook'

SHADOW ON THE SUN is available from:

US: Ingram, Amazon.com, Baker & Taylor, Barnes and Noble, NASCORP, Express Book Machine

UK: Adlibris.com, Amazon.co.uk, Bertrams, Blackwell, The Book Depository, Coutts, Gardners, Mallory International, Paperback Shop, Eden Interactive Ltd, Aphrohead, Waterstones.com

This is a work of fiction. Characters, corporations, institutions and organisations in this novel are the product of the author's imagination, or, if real, are used fictiously without any intent to describe their actual conduct. However, references to real people, institutions and organisations that are documented in footnotes are accurate.

The book is sold subject to the condition that it shall not, by way of trade or otherwise, be lent, re-sold, hired out or otherwise circulated without the publisher's prior written consent in any form of binding or cover other than that in which it is published and without a similar condition including this condition being imposed on the subsequent purchaser.

For more information on the author, the publisher, the book, or other planned books in the trilogy of which SHADOW ON THE SUN is the first, visit:

Blog: http://rjuliancox.wordpress.com
Facebook: www.facebook.com/rjulian.cox
Twitter: https://twitter.com/RJulianCox1
YouTube: http://bit.ly/YBnoah

Author R. Julian Cox

has been a professional writer for most of his life, but SHADOW ON THE SUN is his first novel. He has written a further episode called DEEP EARTH (published March 2015) with a further and concluding part, BRIGHTSTAR scheduled for publishing in 2016.

He began life as a journalist, writing for many UK newspapers and magazines, before moving into public relations and then forming his own publicly listed consultancy, which grew to be one of the largest in the country. Later he formed two magazine publishing companies and was executive chairman of a large, Apple-based computer company. Although originally from York he has lived close to London for many years.

Dedicated to my three daughters
Emma, Eleanor and Amelia.

SHADOW ON THE SUN
Book 1 The Hanford Trilogy

A scientist lies in a coma…

…although appearing dead he is experiencing a realistic dream of another age and of another time. Strangely, in that dream he is witnessing an advanced military aircraft being shot down. It is in the age of King Arthur.

Incredibly remains of that same aircraft are discovered buried beneath a 1500 year old Celtic site on a remote, Cornish beach.

So begins this multi layered novel of entanglement between past and present along with its consequences.

Of a scientific breakthrough gone terribly wrong.

Of an impending environmental catastrophe, of terrorist action and missing computer discs vital in saving millions of American lives. And of a link between past and present in which two lovers have become mysteriously ensnared.

A clock is ticking.

With it a race is on to unravel both aircraft mystery leading to a recovery those computer discs.

But some questions need answering first. And the only person able to provide them currently lies in a coma…

DEEP EARTH
Book 2 The Hanford Trilogy

49 days remain.

A remote British Past and an American Present are still entwined in this stand-alone follow-on to SHADOW ON THE SUN. The deadline is approaching for an imminent project aimed at saving countless American lives.

But just when success seems assured a once thought defunct terrorist cell is again found ready to strike at its heart.

Linked to the project's success so far has been a military aircraft, shot down in the time of King Arthur, but now found to have possible survivors.

A scientist is trying to orchestrate a way back for them from Past into Present. He hopes too to recover a wife who became entangled in a similar fate to the aircraft's and stemming from the same cause.

Can the terrorists be stopped? And in time? Can those trapped in an ancient and hostile history be rescued?

Once again former journalist R Julian Cox displays his deep understanding of the issues and technologies involved. He weaves all of it into the novel's multiple story lines along with evidence of considerable research.

The end result is a thought provoking and compelling read.

Shadow on the Sun

THE "HANFORD" TRILOGY?

This and the next two novels have "Hanford" in the US northwestern state of Washington linked to their plots. This site today suffers awesome contamination problems from which lawyers in the main are getting rich. For they are mired in interminable health and safety arguments with the US Department of Energy - formerly the Atomic Energy Commission – over it's a clean-up efforts resulting from forty years of manufacturing plutonium for use in the country's nuclear weapons inventory. The scale of the contamination left behind and located next to the Columbia River, along with the civil engineering efforts to rectify it, remain unbelievable.

HISTORICAL NOTE

1500 years ago Britain had been abandoned by a Roman army of occupation present for 400 years after Julius Caesar's first invasions of 54/55BC. Behind them they left a fractious and indigenous group of twenty tribes more interested in fighting amongst themselves than fending off new invaders from overseas. Amongst the first of these were the Saxons. Others soon followed.

The Saxons had early success until repelled by a mysterious British figure, possibly of mixed Romano - Celtic descent. According to legend he united the British tribes and led them to a resounding and lasting victory against the Saxons reputedly close to today's city of Bath in Somerset.

Did this figure come from Cornwall, that part of Britain far to the west, where the town of Tintagel has long held a claim to him? Or is it the ancient Somerset town of Glastonbury, once an island surrounded by an inland sea and thus making it the natural choice for the historic Isle of Avalon where he met his end.

Legend attributes the name of this figure to 'Arthur'. But was he real or was he imaginary, simply drummed up by the collective memory of a disparate people who, after the Romans had gone, endured much hardship and later needed a figure, a leader, whom to cherish and respect amidst a sea of trouble?

The literary world has long made much of him with the first Arthurian account written by Sir Thomas Malory in his classic, 'Le Morte d'Arthur' written in the 15th century and later perpetuated by Tennyson's 'Idylls of the King' written nearly four hundred years later.

Throughout the world today the Legend of King Arthur is as alive and flourishing as ever it was. And the debate continues as to whether he ever really existed, or not.

Shadow on the Sun 15

UK: locations of major places mentioned in the text
(not to same scale as USA map on opposite page)

Shadow on the Sun

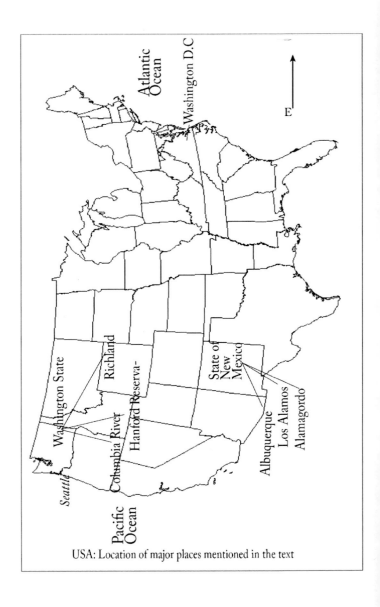

USA: Location of major places mentioned in the text

Star light, star bright,
First star I see tonight
I wish I may, I wish I might,
Have the wish I wish tonight.

(American nineteenth-century children's nursery rhyme.
Anon)

CHAPTER 1

Scientist Dr. Robert Anderson was first to be stretchered off the helicopter. He was one of four bodies being unloaded from the military flight. It had brought them all to a specialist neurological hospital on the outskirts of London from a cliff top lighthouse on a stretch of North Cornish coastline 266 miles to the west.

Anderson was unconscious. So was his wife Katrina, their six year old son, Ben, along with his carer Dolores, a paediatric nurse. They were the only known casualties of the BRIGHTSTAR accident of which Dr. Anderson had been the project's reluctant, scientific pioneer. Soon enough, further casualties would become evident.

For Dr. Anderson his new ordeal was as if "The Light" had trapped him inside a box. Had made him feel neither alive nor dead. Neither warlord, Ar tur nor scientist, Dr Jonathan Anderson. He was lost in a dream, a trance-like state somewhere between consciousness and unconsciousness: locked in a twilight zone. Yet he had been left with an ability to think. And with it, a memory.

That memory behaved like a film projector running an old movie. As Ar tur it allowed him to re run images of the flying object he had seen as it fell into the sea. Allowed him to see the white, inscription stencilled onto a matt, black background:

AURORA VI
High-Altitude Escape Module.
United States Air Force
CXFM130508

And as Jonathan Anderson in a later century he could still see his own equations, the ones leading directly to the object's destruction and to his own, current predicament. He could see much more too and all resulting from his unheeded warnings stemming from his equation's practical application - along with their possible consequences.

Now it was too late. There was a price to be paid. Himself, his wife and son were even now paying it. They had become prisoners, locked in a world somewhere between Past and Present, between who they were and who they had once been. They were all paying the price for his stupidity.

Then like a giant wave crashing onto a distant shoreline, the blackness once more swept over him. Once again he was left drowning. Once again waiting for a rescue that might never come.

He judged it was sometime later. He was hearing voices. They were close and real, discussing his condition. Was he in a hospital?

He tried crying out. He fought to form the words. In his mind he could hear them, but they would not come.

'My name is Jonathan Anderson,' he tried saying. 'Dr Jonathan ... Anderson ... I am a scientist ... and a lay preacher ... in Ripley village where I live.' He tried adding: 'in Surrey ... close to London ...my wife and son live there too. I'm a lay preacher at Mary Magdalena Church.'

He tried repeating, 'My name is Jon... Doctor Jon...My name is Jon.' But it was no use. The words no longer formed in his mind. Instead, he sank back into the abyss. Then he heard voices - the same voices - but this time seeming to come from across an enormous gulf of space and time.

'We're losing him!' he heard a voice say from both somewhere close and also distant.

'Is he there? Is his mind back in the fifth century where I need it to be?' It was an American-sounding voice. How did he know it was American?

Then another voice spoke, carrying with it an accent he mysteriously knew as Spanish.

It answered: 'It is too early. It takes time. His mind has been freed to wander wherever it chooses. We can but hope it is the fifth century; an epoch that for some reason is special to him and is important to us all. We have no option but to wait, to be patient. It could be dangerous for us to make it otherwise … even if we could.'

The American voice spoke again, but this time deeper and more urgent. 'Has he found them yet? Do you think he will find them?' There was added tenseness, almost a pleading.

The Spanish voice answered again, soothingly: 'We do not know. Out technique is experimental. We do not have the data. We've had too few other patients with whom to try our procedure.' He paused then added: 'Did they not tell you?'

'Goddamn it!' said the American voice exasperated. 'There are millions of US lives at stake. Do what you have to, but do it now!'

But the Spanish voice would not be hurried. 'It takes time,' it said again, evenly. 'He's got to have that time. We have to give it to him.'

The American voice would not be stilled. 'He may have it, but I don't! Look, doc, our countrymen are running out of time. They can't wait. Do whatever it takes, but do it now!'

The Spanish voice still refused to be hurried. It remained calm as before in replying: 'We must be patient. We have no choice. He has time beyond comprehension. You need to understand …'

The American voice cut him short: 'All I need you to understand, doc, is that I need what he knows. And I must have it now! What is it you don't understand? How much plainer can I make it? Use whatever drugs you need, but do it now!' The voice had risen to almost a shout.

The Spanish voice continued. 'Real time, imaginary time … they may be all the same to him. Now, he is maybe existing in quantum time. Our time no longer exists for him. All times are the one time. Some scientists call it "block time". What may be eons for us remains no time for him. He is out of our hands.'

As he waited, the man once known as Dr Jonathan Anderson was looking on. His mind had detached from his body. He was looking down on those in the room whilst he floated freely close to the room's ceiling. He saw someone in the room, a woman, come in and then place something on his head; a helmet. It looked grey and metallic with thin wires streaming from its surface.

Mysteriously, from the large room where he was floating, he was able to see across into a smaller room, through walls and across a corridor. Lights on attached equipment in the smaller room flickered endlessly on and off. He became conscious of a low, almost inaudible hum, like that of a dragon moments before it's awakening. He looked closer at other wires and tubes taped to his inert, physical body.

He wanted to shout: 'I am no longer there! I am here! Look up! This is where I am!' But still the words would not come. The force that had possessed him, still paralysed his physical body, would not let him go free.

Suddenly that same force gripped him more firmly, propelled him into a dark void, a tunnel, leading towards a bright and calming light, back to a distant but familiar past, across a line separating Present from Past.

It became deathly quiet.

He opened his eyes. He was back in that other time of Aquae Sulis; that other place of the dream where he could hear the gurgling and bubbling sounds of running water. He felt the heat on his face, could see the steam. But mostly he could feel her, could sense her presence in a way that made him know all over again he had never lost her. She had always been there, in the Present and in the Past, across time, across space.

Had he now become Ar tur? Had it all been a deep and disturbing nightmare?

Was it tomorrow he must fight and die in the way books of the Future already foretold? This Past yet to come could not be altered.

Down those mean streets he knew a man must go who was not himself mean.

Whose 'mean streets?'

Who'd said that? Was it a man called Chandler?

There was no one to answer. All he knew was that he still knew it. That he was this man, with this Destiny.

The dream had slipped him back from the Present to an age 100 years after the Romans had gone from Britain, 1,500 years in the Past; long after Aquae Sulis had bustled with life to a time when Aquae Sulis had become a ruin. With it the dream had become reality.

It was not what mattered. Inside the bathhouse, by the side of the hot pool, she still waited for him. They would meet, up the long flight of steps, through the Doric columned entrance, to where it was warm and dark and silent except for the sound of running water coursing up from a hot, deep earth.

He looked down at his hands, to the amulets she had given him moments before. Small and grey and circular, unlike anything he had seen before.

There was a wish still to make before he flung them into the hot pool. He looked down once more and saw the inscriptions he could no longer read for they were written in a language he no longer understood.

*Property of the United States Government.
If found return to Los Alamos National Laboratory.*

His mind filled with images, of other places he knew, yet places he could never have been; from a Past he had never been a part of yet knew.

It all stemmed from "The Light", brighter than a thousand suns. In that fraction of a second when it had shone it had transformed his mind into something more. Momentarily he was like a child wandering through time and space, like a ghost, a wraith. Katarina, his wife was part of it. She too had experienced "The Light", had somehow become Ganhumara as he had become Ar tur.

Together they had both witnessed "The Light" from the far side of the bay, watched it shine up into the heavens, and seen it intercept the flying object they knew as 'AURORA', then seen it fall out of a cloudless sky into a placid sea.

He knew in this place and in this time he was a military leader; a warlord of the Britons shortly after the Roman Legions had gone, when the Saxons had taken advantage of a lack of defence after a golden age of peace, after he'd already fought campaigns against them, beginning at Badon, and one that must soon end with Camlan.

But the dream would not let him be. In an instant it had moved him on again, so that once again, he must rewitness the same event, but from a different perspective.

Once again, it was later. He was back at the temple of Aquae Sulis, but this time as if seeing Ganhumara for the first time. Yet he had known her forever. It was as if he was a blind man granted the gift of sight, had been woken from a nightmare.

His hand reached out for her as her hand reached out for him. He wanted to feel that familiar, comforting warmth.

They were together, lying naked on a blanket of furs.

It was early morning.

He must go. He must, he told her. She understood as if she had known it forever.

They dressed, then together they walked hand in hand into the bathhouse. His viewpoint switched so that he was seeing it from far above, from a great height he knew to be impossible. He saw too the snow-covered landscape of the surrounding hills. He could feel the cold, could see the fallen walls, the desolation and decay, his army waiting before battle for his last orders. She spoke to him. 'There are so many questions,' she said, 'and too few answers. All I know is that we are here, in this place and in this time. We must act together. There is no other way. It is for the future where people depend upon our actions.'

He knew her words were true, just as they always had been.

'It's you, isn't it? she asked. 'It was always you; my noble Dr Jonathan Anderson.' She smiled. It was a question more than a statement.

The silence filled with the sound of running water. He listened to its symphony before pulling her closer. He could feel the firmness of her body, make out the features of her face, see the violet of her eyes in the dim light.

He looked into them and said, 'You knew from the beginning, but never said, not even after the eons we have already spent together. You left me to realise for myself what it all meant.'

'All that time,' she echoed softly, 'is but the passing of a moment. I would do it again and again for you.'

Tears welled into her eyes. 'I came for you! I volunteered because I loved you,' she said, in a tremulous voice. 'I know it's been hard for you,' she continued. 'It was hard for me too. I had to be certain.' She looked about them before adding,

'Come, we must move from this chamber to be beside the pool.'

He was angry, but it passed as the Future began tumbling back into place in his once-locked memory. She saw it pass from his eyes.

'Do you forgive me?' she asked with the serene tenderness that had captured him from the first moment ... a moment no longer in the Past, but in a distant Future they had once shared.

'We will soon be parted,' she said, 'but it will be for the last time. From then on we will be given eternity.'

'How do you know?' He asked.

'Is not your God, also our God? Did he not say "Love is the greatest of all commandments. None is greater than this?"'

'Yes, yes,' he stammered in reply.

He held her tightly, then kissed her with that longing passed down from both the Past and the far Future.

'There is no more time,' she said. 'Take these.' She gave him the amulets; the two metal objects she had retrieved from 'the man who fell from the skies'. Somehow he already knew of them, as if a distant memory was causing an echo.

'They will protect the future yet to come.' They felt warm, almost alive.

'Cast them into the waters just as the Romans once did, and as was the custom of those before them. We are Romans just as we are Celts. This was their temple. To seal the Future we must do the same.' She said it softly as her hands enclosed his that now held the discs.

'The waters of the gods will protect them until the time is right.' She removed her hands, urging him on. 'Take them. Throw them. Then together we will make our wish.'

He looked at her, at the hot pool at their feet, then at the objects in his hand. He saw the glinting of the many coins in the pool and at the jewellery thrown there by generations who had gone before.

In the flickering flames of the torches he saw again the inscription he could no longer understand:

Property of the United States Government.
If found return to Los Alamos National Laboratory.
Shipping address: PO Box 1633, NM 87545, USA.
Serial No: 175890/1603/11.
Possession is an offence without proper authorization.

Ganhumara looked at him. Time was short. They were locked into the endless loops of time with no beginning and no end.

The warlord drew back his arm. He turned to her with a puzzled look on his grizzled face. 'We are being watched,' he murmured.

'I felt it too, ever since Ynys Witrin,' she replied.

There was no more time. He threw the amulets high into the air and watched them fall into the pool. They made their wish. Then they turned, hand in hand and walked along the side of the hot pool, past the other chambers, and out into a frosty night filled with billions of stars. As they saw them a ripple seemed to pass across the sky.

With it, they were both plunged into an infinite void.

Six month earlier…

CHAPTER 2

That year's early British weather had been cool with an initial, wet spring followed by a long, hot summer stretching from April into September. That part of the southwestern UK called Cornwall, a peninsula jutting 200 miles out into the North Atlantic, had been hotter than usual. Even normally its climate did well by benefitting from the warm current of the Atlantic Gulf Stream as it flowed gently up from the equator.

For the English Southern Cornwall was often referred to as 'The English Riviera,' a deliberate comparison with the far hotter and more glamorous 'French Riviera', with its cities of Canne, Nice and St Tropez located 700 miles away along the French Mediterranean coast. Here the annual Palme d'Or film festival with its attendant flocks of sun bronzed people had for long provided a spotlight on the world stage. But by comparison the less well known southern Cornwall still faired well with it's sun shine quota which was generally better than the north where it was often colder and wetter.

It was during this Summer period that a group of young archaeology students from Exeter University, came to know North Cornwall's Pendragon Point well, and its close-by Pendragon Quay. The city of Exeter situated half way along the peninsula and close to its southern coast had long laid claim to being the most southwesterly of Roman cities fortified during the Roman invasion of Britain during the early part of the first millennium, almost 2000 years ago.

For several months the archaeologists had worked and slept nearby while excavating a long-suspected Celtic site dating back to AD 491. It was the period of Britain's greatest king, King Arthur, some had noted. Others, sceptically, voiced a belief that he was entirely fictitious, only ever having existed on the pages of classic Western literature, particularly with Sir Thomas Malory's Le Morte d'Arthur, written in Newgate jail during the fifteenth century.

Before the 'dig' and for student information purposes, their university faculty had produced a short leaflet describing how the broad headland near where the 'dig' was located had first been formed by volcanic activity, at a time when the British Isles had lain south of the equator. Most of it had then been submerged beneath a shallow tropical sea. But 400 million years of movement by the earth's tectonic plates had caused it to drift 3,000 miles north, along with the rest of the British Isles.

The student's leaflet included a map showing that ten miles across the bay of which Pendragon Point and Quay formed a part, was the village of Morwenstow, along with its cliff top cluster of white-painted, circular radomes and slender radar masts, belonging to a UK Government satellite listening station. The southernmost end of the bay was punctuated by the island of Tintagel that still clung to the mainland by a narrow isthmus. On each of it's sides steep cliffs dropped away towards a sometimes azure sea. Over millennia, constant wave action had chiselled the deep and mysterious Merlin's Cave, while on the island itself local legend had it that this was where the fabled King Arthur had once held court.

The Exeter students had come to know the narrow, twisting road connecting Pendragon Point, and its associated but long-disused lighthouse, with Pendragon Quay seven miles away. As the road approached the quay it descended precipitously, before ending in a small, cobbled area with its own tiny car park, capable of holding no more than five or six cars. To one side, a long series of low stone buildings were separated from the facing Smuggler's Inn by a twenty-five-foot wide paved thoroughfare. Twenty yards beyond the buildings the lane ended at a low stone wall, beyond which lay a semi-circular bay a quarter of a mile in diameter. Stratified granite cliffs rose vertically, while at their foot a strip of sand fifty feet wide marked a small beach, left mostly uncovered except by abnormal spring tides.

Low tides usually revealed a pebbled sandy expanse, littered with fallen rock and the occasional house-sized boulder. Further out in the surf, islands of fallen granite gave further evidence of the sea's remorseless power, even on the toughest of materials.

High up on the usually dry sand-covered part of the beach, closest to the cliffs, the Exeter students had already been working for several hours in the hot sunshine. The site was enclosed within a fenced area measuring twenty-five feet by ten, and reached a depth of thirteen feet, meaning the sides were well-shored with timbers to guard against possible collapse, and even possible immersion by a renegade and unexpected high tide. These tides, locals 'in the know' had advised, were always possible and had to be constantly guarded against.

The oldest and most experienced student, a postgraduate in her mid-twenties, had been placed temporarily in charge. Tomorrow would see the start of the long university holidays, already heralded when their associate professor had left the previous day to take another party of students on a six-week dig to Troy in Hissarlik, Turkey, south-east of the Dardanelles.

Each student had been allocated their dedicated task. All were involved in carefully scraping away the lower strata of soft, brown clay that lay beneath the sand. Each knew it was the last day; the last opportunity for 'discovering' anything important. The professor's brief had been to try and find anything they might have overlooked previously. But most important was to leave the site 'tidy', and as safe as possible for several months.

From previously uncovered artefacts the notion had already been advanced that the site had once been a Celtic smelter; one of many in that part of the world, resulting from the area's rich abundance of tin, copper and silver, and some gold. Its remote location further suggested it could have been an armoury because of the ancient weaponry already found. Usually, such armouries were hidden from normal tribal activity, for the skills of those who worked there were highly prized and had to be protected from marauding warriors for whom 'kidnap' and 'forced labour' were a way of life and survival.

As with most archaeological sites, this one had been terraced as the dig had got deeper. Spurred both by enthusiasm and youth, the students were working diligently. Their 'supervisor' was inspecting a shard of red earth pottery she thought had originated in Greece, when an excited shout came from a student working at the deepest level. She looked over and saw his arm wave. She walked over. Probably a false alarm, just as there had been many times before. But this time she immediately saw his cause of concern, and was equally puzzled by what she saw. She scrambled down to make a closer inspection.

Below the mud was something she had never seen before: a black, scorched and mud-streaked artefact, the shape of which she had never previously encountered. It was immediately apparent to her that it was something large and curved. The student who had discovered it raised his excavator's trowel and brought it down fearlessly, wooden handle first, on to the object below. There was a dull, non-metallic thud. The two looked at each other quizzically. Neither said anything. The timbre suggested something big, very big, and hollow.

There was an unmistakable emblem stencilled on to its smeared and burned surface that, irrationally, had the patina of age. What they both saw was impossible. There was no way that insignia could lie beneath what had already been discovered, and which carried proof of great age, maybe going as far back as two thousand years.

They continued to stare in disbelief. Others from the site joined them. At first no one spoke. Then everyone began to speak together. She asked them to be quiet while she thought. She could deduce from its overall shape that it was a craft of some kind. She placed one hand on it. It was not cold to the touch, and felt like some kind of plastic. That, too, was not possible. But that was the least of the most impossible possibilities; it was the insignia that drew everyone's attention.

It was a 6 x 4-inch flag of the United States, beneath which were the white, stencilled, letters and numerals on a dull, black surface:

AURORA VI
High-Altitude Escape Module.
United States Air Force
CXFM130508

CHAPTER 3

There had been a thin covering of snow with the weather unusually cold for March, well below the historical, monthly average of 40°F. It was almost noon, Pacific Daylight Time, as the reporter was being driven round what was left of the small town of White Bluffs. It lay not far from the Columbia River almost in the middle of Washington State, itself situated in the North Western part of the United States immediately below the Canadian border.

It was the last day of *The Washington Post* journalist's four-day trip. He had already visited the coastal city of Seattle, famous for being home of software company Microsoft and plane maker, Boeing before driving 200 miles almost due east to the city of Richland. Then it had been a further twenty-five-mile drive to that stretch of the Columbia River downstream of Priest Rapids Dam, to the 600- square-mile area encompassing the former Indian reservation of Hanford.

Yesterday it had been the turn of the Department of Energy to be interviewed and then, later, it was major construction contractor, Bechtel. Although the PR people were seemingly warm and friendly he knew they were all on their guard, nervous, with their eyes darting everywhere. He had been nervous too.

He knew already he had the kind of story his section editor was after. All that remained was this one, last further element. The editor had said before he'd gone that he wanted a 'fresh angle', a 'human angle' that showed through despite all the impressive technology and buildings he felt sure he would encounter. The journalist knew more than anything that he had to surpass what *USA Today* had already published. He thought the extra mile he had just embarked on would clinch it, would add the all important 'colour' he needed.

'It just beggars belief,' he said at last, in that peculiar accent his Native American Indian guide had noted in the short time they had been together. The journalist was British, and had only been with the Post a short time. The old man driving the 4 x 4 was from the Yakama tribe, one of twenty-nine tribes in the area. He did not reply to the journalist's comment, so the statement continued to hang lazily in the air.

The Indian's father and grandfather had told him the stories of how the men had come to escort them out: how in the early 1940s a huge manufacturing complex had sprouted up on land that used to be theirs; how nothing had been done to replace their former way of life. And in the long time that had passed since then, he and others like him had become used to showing the curious around what was left.

'So this is the human cost of Einstein's famous $E = mc^2$ equation?' said the reporter. 'All this desolation, just to set up the "B" reactor and all the other plutonium - and uranium - processing facilities?'

'We have learned to live with it,' said the Yakama Indian at last. 'They came and took away from us the lands of our fathers and our fathers' fathers. Before that, and for ten thousand years, we had hunted and fished here. Then they took it away and left behind a wasteland.'

They drove on through the remains of the once-thriving town of White Bluffs. Back in the early part of the twentieth century it had been home to a population of 900, when a railroad called 'Sagebrush Annie' had carried riders backwards and forwards to and from another small town called Hanford. It had been named after Cornelius Hanford, a well-known area judge of the time.

What the reporter now saw, instead of bustling commerce, was a desolate mixture of flat land and rubble for as far as the eye could see, except for two gaunt shells of buildings seemingly deliberately left standing.

One had been the town's two-storey high school, constructed in 1916 and once regarded as one of the grandest high schools of its time. The roof had caved in and some of its walls reduced to rubble by US Army bombing practice. The other building had been the town's bank. It had once stood on a corner where two thriving high streets had met, where shoppers had jostled with each other among the rows of bustling, family-run businesses.

Now there was nothing. Not since 1943, when the government had given the inhabitants thirty days to get out. In exchange for a little money they had gone for they were given no other option. The Native American tribes had then been banned from their ancestral homes and gravesites. There was a war to fight, they had been told, and 'the land was now needed for important war work'.

With the inhabitants' departure the twin towns had died to become the first step in what would become a vast, ecological disaster. In exchange would be the building of an incredible stockpile of 60,000 nuclear weapons. Ironically, most later destroyed as part of the Cold War's long-time-coming 'peace dividend'. It would mean White Bluffs and Hanford had ultimately died for nothing.

The two men drove on in their 4 x 4, past dumps of day-glo-painted oil drums piled high and rusting. Many lay on their side with some leaking their contents so they stained other drums or left bright orange patches on the earth.

They passed a single, yellow sign proclaiming forlornly: 'Danger! Radiation! Hazardous Waste'. Occasionally they would pass a contractor's truck carrying workers either to or from other locations on a vast site someone had infamously labelled 'the most radioactively contaminated place on earth'.

'Our tribe has lived here for ten thousand years,' the old man repeated sadly as he drove. 'The summers were pleasant and the winters mild. We hunted game. Others farmed the scrubland. Our people died here, and we buried them with their ancestors alongside the river.'

Eventually he pulled the 4 x 4 up on a steep slope alongside the muddy, dirt track. From there they could look out over the Columbia coursing its way, many hundreds of feet below and five miles distant across the plain. In a low monotone the Indian said: 'You see to the right hand side of the river? That's K Basin West. The one on the other side is K Basin East. They were reactors and short-term storage sites. For some reason they built them within 400 yards of the river. They said each had a design lifespan of maybe twenty years. During that time they would be safe.' The Indian looked forlorn. 'But that was eighty years ago,' he added. 'Today they leak their contents into the ground. It then soaks into the groundwater and into the river killing the salmon.

'No one dares go into K Basin East anymore. The radiation levels will kill a man in minutes. The other reactor site is safer.' The old man pointed with his bony finger. 'In there they store 90,000 drums of radioactive waste. They are kept in large tanks of water to keep them cool and stop them exploding from a gas build up.'

But the reporter wasn't listening any more. He'd jumped out of the car and was busy taking photographs of the faraway buildings. A few minutes later he climbed back in.

The old man continued. 'We have everything here, a real garbage site. 'Over there are forty spent nuclear reactors from submarines and other navy vessels with nowhere to go. See that pile over there?' He pointed. 'Thousands of radioactive fuel rods from power stations. And over there is waste from plutonium production and uranium enrichment. More keeps coming.' He stopped and sighed. 'There are another 128 sites in America just like this one. But Hanford is the worst.'

'I think the American people should know more about it,' said the reporter. More about what we've done and the environmental damage we've caused. They should know, so we can clean it up before it's too late.'

The old man chuckled as they drove off. Who would listen? He silently thought. 'C'mon, son. I'll take you down to the river, to where the sockeye salmon once bred. See what you make of that!'

'Have we got enough time before the flight back to Washington DC?' the reporter asked.

'Sure,' the Native American said almost immediately before checking his watch. '12.40,' he added. 'Sure there's enough time.'

∞

The following day, the streets of London were cold and swept with rain as Dr Jonathan Anderson walked down the three steps leading into the small Italian restaurant. Anderson had a problem. It was a professional problem, and one that over many months he felt conflicted increasingly with his religious faith. Sooner or later he had always felt that he had to talk it over with the only person he could trust.

Sheltered by the restaurant's red-brick porch was a freestanding chalk board on which had been hand-scribbled that day's lunchtime menu.

'Hi Jonathan! Over here,' a voice shouted in greeting as soon as he pushed through into the crowded interior. A long, thin but athletic-looking man with fair hair in his mid-forties was half- rising from his chair. Jonathan walked over, shaking the drops of rain from his short jacket before handing it to the restaurant's receptionist.

'God, it's so crowded and so noisy,' said Jonathan to his elder brother, Edward. 'Couldn't we have gone somewhere quieter?' he added in quick mock criticism.

'There's a big case on over the road,' retorted his brother. 'You must have seen it in the papers or on TV. A lord of the realm kills his wife, then tries to cover up the dastardly deed by burning down the marital home and alleging it was 'burglars wot done it' … besides, if I'm picking up the bill, which I assume I am, then I get to choose where we go. So shut up and sit down.'

Jonathan meekly did as he was told. He knew the way it was in that part of London. The restaurant, the Stromboli, was down Little Sussex Street, opposite the huge neo-Gothic bulk of the Royal Courts of Justice.

'You know the form,' Edward continued. 'This place is our unofficial canteen.'

'I remember from the times before,' said Jonathan. Whereas he was dressed casually in jeans and open-necked shirt, Edward was wearing a three-piece grey pinstripe over the top of a collarless shirt. Others seated around them were similarly dressed. The atmosphere was noisy and good-humoured. Edward could only have been one of them: a barrister, someone skilled at presenting the guilty or innocent parties case in a court of law.

'So you know the food's good. I recommend the lobster, and for starters there's the whitebait. But, of course, choose what you like. I told you before you came I wouldn't have much time before I was due back in court.'

'Defending or prosecuting?' asked Jonathan out of mild curiosity.

'Defending,' replied Edward.

'Whitebait and lobster will do fine for me,' replied Jonathan.

'Okay. And for me it's the steak, the *bistecca alla fiorentina*.' He called the waiter over and placed his order, including a bottle of wine.

'We'll have "white" in deference to yourself,' said Edward. '*A Pinot Bianco Adige*. It's an intense, citrusy wine with a note of lemon. You'll like it.'

Edward's chambers were one of many clustered around the Courts of Justice that housed some of the best-known legal firms in the country. Most of those offices, or 'chambers', were located around a green and leafy square collectively known as The Temple. It housed two of the four 'Inns of Court', the Inner and the Middle Temple, that could trace their origins back to 1322. To become a barrister one had to belong to one of these four 'Inns of Court'. Along with the Old Bailey, a short walk away up Ludgate Hill, the 'RCJ' handled the country's most famous trials, which was why outside its main entrance there were always door stepping journalists.

At 43, several years older than Jonathan, Cambridge-educated Edward was a rising star within his chamber of eighty barristers and clerks, all headed by six legally qualified knights of the realm. Jonathan had passed their offices as he'd walked to the restaurant. Every time he walked passed, he always noted who'd come and who'd gone from the names listed on the large cream-coloured board displayed prominently outside.

'We've got to press on, Jonny. So I've already looked into the background of what you've asked me. It doesn't look good. I've not professionally handled a case involving the Official Secrets Act, but one or two others in our chambers have. But they only move if 'fees' are involved. Trust me, you don't want to go there. Not at present anyway. A million quid doesn't go far with them.'

A waiter came across with the bottle of wine and began uncorking it. He asked Edward to taste. He did so, approved, and then stopped talking until the waiter had poured the drinks and moved away.

'Now I remember warning you about this right at the beginning, five years ago when you started. The Official Secrets Act is not a contract of employment, although you're normally asked to sign it. It's actually the law of the land, and whether you sign it or not is irrelevant.' He took a sip from his wine glass. 'Gosh, that's jolly good stuff. What do you think about it, Jonny?'

Jonathan followed suit. 'Better than we get in our staff canteen,' he murmured.

'Yeah, it costs a lot more too, I would think.' Edward placed his glass back down on the table. 'So you're bound by it. They've got you by the short and curlies because the Act applies specifically to Crown employees or contractors. Which is what you are.'

'The original Act of 1911 was brought up to date by the 1989 legislation that replaced Section 2. Where it applies to you, Jonny, is it removed any "public interest defence". So if you reveal something that came your way by virtue of your employment, particularly if working on a classified defence project, you're banged to rights. You can't go crying to newspapers – which I have the feeling is something you've been thinking of doing – because the government is doing something of which you no longer approve. End of story. The penalty could be two years in the slammer. There's no defence.'

'That doesn't sound too good, Ed,' said Jonathan, almost apologetically. 'The question then is, what do I do?'

'You can resign. But if you do, then keep your trap shut. That's my advice in a nutshell.'

The waiter came back over, carrying their food order on a tray.

'Thanks, Franco,' said Edward.

'How sure are you about your work being dangerous?' he asked.

'Pretty sure,' replied Jonathan. 'I've been working on it for months. Stemmed from new data I got from CERN, you know, the …'

'… Yeah, I know who they are,' said Edward impatiently. 'The European Centre for Atomic Research in Geneva!'

'Yeah, they're the ones. They'd done the research for another area they were working on, but I saw it could have application in mine.'

'And you're telling me now that the intellectual rights the government so generously paid you such an awful lot of money for, are potentially waste paper ..? He thought about it for a few more moments before adding '… though don't think there's much they can do about it in practice. Caveat emptor, and all that.'

'No, Edward. I'm not saying it won't work. Listen up. The Project'll work fine. It's the other things I've belatedly discovered it could do that bother me –' Jonathan paused, and then added for emphasis, '– they now bother me a lot, so much so I think they're against the tenets of my religious faith.'

'I'm a simple lawyer, Jonny, religious faith is outside my area. And besides, I don't want you telling me too much about what it is you're actually working on right now. Though from when I last advised you, when you originally took the government's cash, we filled in the client agreement forms so we had lawyer/client privilege. But that was four or five years back. I'm now advising you, just as your brother. From what I remember then, your intellectual property was to do with a neat way of achieving "fusion", the energy that powers the sun. And this has been incorporated into some sci-fi beam weapon that can shoot down incoming ballistic missiles. Now you're telling me there's a problem. If you like, a shadow on the sun.' He took a leisurely further draft from his wine glass, temporarily feeling smug at his play on words. He then looked Jonathan in the eye before saying seriously: 'You're telling me it could be dangerous?'

'It could be. The emphasis is on the word "could".'

'How?'

'At worst, it could fracture the space-time continuum. God knows what the effects of that could be. At its least it could affect people caught in its beam.

'I don't really know the consequences of that. Not my field, but if the brain works at a quantum level, as some think, then ...' He shrugged his shoulders. 'It's never been done. And when it is, it could be too late ..!'

'Let me ask the obvious question,' Edward interjected. 'You've tried telling the person in overall charge. Sir Jack Geisner, is it?'

'Sure I've told him. But he's not interested. They think like you do, that as the moment of truth approaches, scientists like me get cold feet ... we take our money and run.'

'The problem is I can't really prove it except by some complex mathematics only six people in the world would understand.'

'And you can't discuss it with any of them because of the Official Secrets Act,' added Edward.

'That's it in one,' agreed Jonathan. 'It's Catch 22.'

After he'd tucked into a good portion of his steak and quaffed a further large amount of wine, Edward said: 'I think you're stuffed, son. All I can say that might be of help is that the government have been highly unsuccessful in pursuing those few cases they've brought under the Act. Mostly, and at the last moment, they pull out. They fear prosecution will mean having to disclose more information than they're comfortable with. So the Act actually becomes self-defeating. It's a view, but not one I'd advise you to accept.'

Edward turned his attention back to his steak before noticing Jonathan had hardly touched his lobster. 'Go on, eat up,' he said. 'Too good to waste. The condemned man ate a hearty meal, and all that.' He roared with laughter before continuing eating, and then asked: 'So when does the Project begin roll-out?'

'It's soon, Edward. Very soon. It's a prospect that's giving me nightmares.'

'You're going to have to live with them. Contractual obligations to one side, employers can hardly ever hold you if you decide to go. My advice still stands. Resign and keep stum. It's true you've already received a big lump sum from them as the basis for you joining … is there any of it left?'

'Most of it,' replied Jonathan.

'Good. Good. So you don't need any more of their money. At least not for some time. Good position to be in.'

'And Katrina is doing very well with her television work,' added Jonathan. 'That's why the money has held out so well.'

'Excellent.' Edward took another large draught out of his wine glass then, noticing it was almost empty, began refilling it. Belatedly, he asked Jonathan: 'Do you want another?' Without waiting for an answer, he began refilling Jonathan's glass.

'So that's the business out of the way, Jonny.' He checked his gold wristwatch. 'I've got to go shortly. But quickly, tell me, what else has been happening in your life? It's over a month since we last met. Tell me about Katrina. What she's up to, and how is your son Ben faring, poor bugger?'

∞

As Jonathan and Edward were completing their lunch it was 7.30 a.m. Eastern Standard Time in Washington DC.

'Mr President,' the woman and man said in deferential greeting. They then stood up in unison as President Juan Sanchez walked into the Roosevelt Room directly opposite the Oval Office of the White House.

The woman was dressed in a sombre, dark blue trouser suit, the man in ill-fitting slacks and jacket stretched over an overweight frame. The president's Chief of Staff, Cliff Wurzburg, was directly behind the President. Sanchez motioned the two to sit as he withdrew a chair from beneath the long, reproduction Queen Anne table, capable of seating sixteen people, as it did when the White House cabinet met, usually once a month.

'Sorry we're late, gentlemen, but the president is on a tight schedule so we've gotta be quick,' said Wurzburg. Now the president and I both know you, Larry, from heading our Department of the Interior. But your associate?'

'Sorry, Cliff. This is Carla Offendorf from our US Geological Survey. It's her guys who've been doing the checking for you.'

'Okay, Carla,' said the president sympathetically. 'So tell us what you've got?'

'Mr President, you have a good guy working for you in the form of Professor Robert Pahlavi. He's from Iran, and a world-class authority. A lot of them are from that part of the world. Their own country is so wracked with earthquakes. What he's told you checks out so far, as we can tell.'

The president looked thoughtful but said nothing. Clearly he was expecting more. Carla Offendorf responded, 'We all know that whole area is tectonically unstable. The plates extend well to the east from the Pacific coastline. It's the northern end of the San Andreas. The Cascades, in particular Mount St Helens, has been rumbling for centuries. So when I tell you it could go at any time, that's not news. The question is, when? This is Professor Pahlavi's sphere of specialisation.'

She looked for signs of understanding from her audience. There was none.

'Yeah, yeah,' said Wurzburg impatiently. 'What you've told us so far the whole world knows. The question is, what is the scheme Pahlavi has to protect Hanford site? To put it bluntly, is it safe? Or should he be put away with the Looney Tunes folk?'

'We've studied our maps of the area, overlaid them with our charts of what we know about the geology, done computer simulations to refine our best guess …'

'Jeez!' Exclaimed Wurzburg. 'We're paying you for "best guess", as you call it.'

'Cliff, understanding earthquakes is not easy. If it was, we'd have sorted San Francisco out years ago,' said a mildly irritated Larry Schwarz. 'You've gotta be certain. Can't move millions of people and then tell 'em later it was all a mistake. They can all go back!'

'So what do we do?' asked a still ice-cool President Sanchez. 'Professor Pahlavi says we have fourteen months. Maybe eighteen at most. Eighteen months is when it hits, so we've got to be well ready before then. That gives us maybe a year from now. About next March.'

'The whole area is certainly more active now than it's ever been. The computer simulations suggest your plan to contain Hanford will work …'

'At last,' said Wurzburg. 'That's what we wanted to hear.'

'Yes it is,' said the president. 'Now I'm sure I don't have to remind either of you that this is highly classified. What we've talked about doesn't leave this room. Am I clear?'

'You're clear, Mr President.'

'Good,' he responded. 'Now we've gotta go. Thanks for coming over to see us.' The president rose from his chair and Wurzburg followed suit. 'Many thanks,' said Sanchez as he exited. A slim woman entered as they left, and held the door ajar in expectation of escorting the man and woman out.

∞

Later that same evening, in the small Surrey village of Ripley outside London, Dr Jonathan Anderson, along with the vicar, had just finished shaking hands with their departing and small congregation. They made up the few who had earlier made their way to the weekly evensong service.

Jonathan and the vicar were both standing on the steps of the village's twelfth-century Mary Magdalena church situated just off the high street. For the past two years Jonathan had been the church's lay preacher, and during that time had been working hard building up evensong attendances. His success was measured in ones and twos rather than a stream.

His role was voluntary and part-time, alongside his professional life as a talented mathematical physicist whose original insight had led to his appointment as principal architect of the revolutionary and still secret BRIGHTSTAR project. His aim while at Cambridge had been to develop a limitless energy source for a world progressively running short of fossil fuels. But someone in government had understood that his work could have application within the defence industry. They'd offered him more money than he'd ever dreamed of to switch roles, and at a time when he and his wife Katrina needed it to support their mentally handicapped son Ben, who was barely six years old.

Recently Jonathan had completed a reworking of his original theory, a fuller assessment of his original work, much as Einstein's Special Theory of Relativity, published in 1906, had been an adjunct to his later General Theory published in its final form years later.

His new work had revealed unexpected results in certain circumstances that, in his view, were too important to ignore. At worst they meant that after all the work already done, the hundreds of millions invested, the livelihoods of thousands of people employed, BRIGHTSTAR could be a defence dead-end. Within the next few days he planned to tell the Project's leader of his decision to resign. To continue was now against his religious faith. The risks were too big. The only way out was for him to accept his brother's advice. But before he did resign, there was one last option open, a way of finally validating what he now thought, but importantly without violating the government's Secrecy Act. He still hoped against hope it wouldn't come to it.

That night, as Ripley's villagers found their way home, and with the earlier discussion of the day with his brother Edward still lingering, 'resignation' was still the major issue on Anderson's mind.

Behind him on the church steps, as the last few members of the congregation shook his hand, was the vicar. They were all chatting amiably among themselves as if they had all the time in the world.

Anderson and his wife Katrina had been living in the village for two years. Apart from being eighteen miles from the centre of London their choice was more to do with Ripley's closeness to the Atkinson Morley Neurological Hospital in Wimbledon that their son needed to attend on a regular basis, plus Ripley provided relatively easy access to their respective locations of employment.

Each day Jonathan travelled across country to the British Atomic Weapons Management Company based on the far side of Reading in Burghfield, while his wife drove most days to Reading University where she had been appointed a Senior Research Fellow in Ancient History, funded in part by the Royal Society. During the last three years she had become popular with film production companies, who liked her 'take' on ancient history and its impact on modern life. And they were prepared to pay well for it.

The vicar had turned to re-enter the church and lock up for the night. The last of the congregation had departed. As he did he held Anderson's arm and said gently to him, 'It's okay, Jonny, I'll lock up. You can be on your way home.'

Jonathan watched him go, locking the church door behind him, before he remembered he'd forgotten to mention his 'hot news'. He shouted through the letterbox: 'Forgot to mention. Been accepted for theological college.' He waited to see if he'd been heard but there was no response. He turned and walked off along the High Street, relying only on his Church of England black cassock and white surplice to protect him from the cold night air.

Anderson was a large and well-built man in his early thirties with dark and closely cropped hair, a narrow face with a mouth that carried an easy smile that had become familiar in the locality as he went about his pastoral duties. A few minutes later he was shouting 'Hi darling,' as he entered the front door of their Victorian cottage a short walking distance from the church steps. Most notable in the recent and extensive house restorations were the wider than usual external and interior doors, with a lift going up to the upper floor of the new extension containing two bedrooms and en suite shower rooms. It had the hallmark of alterations tailored to suit someone with mobility issues.

'Hi, Jonny,' came the answering shout of his wife. 'I'm in the kitchen. Working on my new TV script.'

Once in the kitchen Jonathan walked over and kissed her on the cheek. 'Hi, Mr Science Man,' she said as she sat at a large oak table bearing the scars of many past battles. Books were spread all over its surface and around her feet. Jonathan pulled up a chair and sat opposite. He was tired.

'Weren't you cold going out in just those clothes?' she said.

'I was fine. Don't worry about it.'

'But I do, but I do,' she responded. 'Can't let the country's top scientist catch his death of cold.'

Jonathan laughed a light laugh that didn't fool her.

'Tough evensong?' she asked, without looking up and with a trace of a Welsh accent.

'Nope. Piece of cake. The vicar was there to help me. He's doing it on his own tomorrow.'

'Oh?' She looked up.

'Yep. You know it's tomorrow night I'm due over at the site for the lecture. My old tutor from Cambridge is giving it.'

'Ah, yes,' she said, remembering. 'Let me look at our wall chart.' She put her books down and went over to study it where it hung on the wall. She lightly kissed Jonathan as she passed. 'Ah, yes. It's down here. You can go. Dolores won't be looking after Ben. It's her night off. I'm doing it. And I'm back here about four-thirty after visiting the Royal Institution in Albermarle Street, who want to make sure they're getting their money's worth out of me.' She turned round and went back to her seat. 'So everything is A-OK.'

Katrina Anderson was a striking woman. Thin, long black hair, pale white skin, thin face, five feet ten inches and set off with rare, violet eyes that had captured Jonathan's heart from the first moment he had seen them when both were students at Cambridge. She had long, slender hands too, more redolent of an artist or a piano player. Maybe even a violinist?

'It's a big night for you tomorrow?' she asked tentatively.

'If I do it right, my old lecturer may validate what I already know, and without me risking breaking the Official Secrets Act. Without me asking him straight out.'

'Okay then. We're all set. But changing the subject for a moment, listen to this. I was reading this book just before you got back. I think it's rather good. I'm hoping to use it as part of my next TV series. The author's dead now, and his reputation as a historian has suffered from what he's written right in this book called *The Age of Arthur*.[1] She held it up for Jonathan to see its cover.

'It's about King Arthur, his life and times and what happened afterwards. I'm looking at it as my next film project. It says in the preface "it is the classic account of the British Isles from the fourth to the seventh century". This is what he says at the end of his very thick book. Now listen up:

> [1]*Therein lies the importance of the short-lived realm that Arthur salvaged from the ruins of Rome. The tales that immortalised his name are more than a curiosity of Celtic legend. Their imagery illuminates an essential truth. In his own day Arthur failed, and left behind him hope unfulfilled. But the measure of any man lies not in his own lifetime, but in what he enables his successors to achieve. The history of the British Isles is funnelled through the critical years of Arthur's power and its destruction, for thence came the modern nations. The age of Arthur is the foundation of British history; and it lies in the mainstream of European experience.*

She looked at him over the top of her heavy-framed spectacles. 'Good stuff, huh? I'm gonna use it in my programme, and screw what the academics think afterwards.'

'You gotta do what you gotta do,' said Jonathan back to her. 'It sounds like his stuff is written from the heart.'

'You bet,' she said, then adding, 'And you should lead with your heart over your BRIGHTSTAR stuff! We don't need the government's cash any more. Nor does Ben. Tell 'em all to go to Hell.'

'I might after tomorrow, after the lecture,' replied Jonathan.

---

[1] *The Age of Arthur, published 1995 by Phoenix, a division of Orion Books.*

CHAPTER 4

It was 6.45 p.m. GMT at the vast and sprawling complex that was the British Atomic Weapons Management (BAWM) site situated outside the Berkshire town of Reading, some forty miles south-west of London. It comprised two sites, each of over a thousand acres apiece. The larger was Aldermaston, with the smaller at nearby Burghfield. Although their two functions overlapped, the former was responsible for implementing nuclear warhead upgrades and maintenance on the nation's stockpile, while the latter spearheaded nuclear weapons research and development. Recently added to Burghfield's responsibilities was a new site at Morwenstow in northern Cornwall, home of a former GCHQ satellite listening station and currently nearing completion for a new role.

GCHQ was the British Government's vast and highly secretive organisation for code breaking and electronic surveillance compared in function to the similar American NSA organisation. It was because of its secretive nature that GCHQ Bude, as the Morwenstow site was known, seemed to be an obvious candidate for the new duties it would soon have to undertake. Part of its new role encompassed a satellite site at the ancient town of Glastonbury, one hundred miles to the west in the county of Somerset.

Aldermaston had started life as a Second World War RAF airfield and Spitfire aircraft production facility shadowing the main and frequently bombed production plant at Castle Bromwich close to Birmingham. Manufacturing the famous single-engined fighter was a small beginning compared to what it had become. Altogether it now employed 4,500 staff, along with 2,000 contractors and subcontractors.

Normally the administrative and executive offices at Burghfield shut smartly at 5.30 p.m. each evening, when the staff finished their day's duties. But that evening something special was happening so the lights in many of the office blocks were still burning. It heralded the last of a series of lectures the company had been running in which major scientific figures had been invited to talk on their area of specialisation close to but not linked to anything BAWM were engaged in. There had been talks on the 'Higgs boson' and its associated 'Higgs Field' from someone at CERN; on black holes and holograms from a member of the Royal Astronomical Society, and on developments in string theory from someone at Canada's Perimeter Institute. There had been five such lectures altogether and all had been well attended.

That evening in the company's largest lecture theatre, more prosaically known as room 259 in office block H67, a man who had once been a leading authority in quantum physics was speaking on 'religion'. His specific topic was: 'A providential aspect of the universe cannot be ruled out'. It was deliberately chosen to be controversial, as science and religion were often seen as the antithesis of each other. But the guest speaker had a foot in both camps. For as well as working with some of the greatest names in quantum mechanics, later in life, at the age of forty-two, he had chosen to leave his subject to train as an ordained Anglican priest.

∞

'Go! Go on! Leave! And leave right now! You're going to be late unless you go this very minute!' Katrina herself had only just got back home after her meeting at the Royal Institution in London, and had been surprised to find her husband still in their home.

'I came back to see Ben, to kiss him goodnight,' spluttered Jonathan.

'I know why you came back,' said his exasperated wife. 'It was thoughtful of you. But you really must go.' She pushed him towards the front doorway of their house. 'It was madness to drive over from your office and then have to drive all the way back again so you could attend the very seminar that means so much to you.' He saw the pleading look on her face and knew she was right. 'If you don't go now, and I mean right now,' she continued, 'you'll miss it …'

'It's the mathematics,' he blurted out to her. 'It was all in the mathematics,' he repeated. 'I couldn't ignore what they were saying to me over and above what they said about the Project. Tonight will clarify everything …'

'I know,' she said. 'We've talked about it. Often. Now you have to go.'

She kissed him. 'I love you,' she said, 'whatever happens.' She kissed him again, passionately. Jonathan felt warm. He was ready for what lay ahead. He grabbed his coat from near the door, turned, and was out and hurrying towards his parked car. She followed him and kissed him again and felt his tight embrace that warmed her. Then he was in his car and away.

She watched him go. Tears crept into the corners of her eyes. 'Idiot,' she said underneath her breath.

She'd always understood the bargain he'd made with the devil. She'd understood why he'd accepted the government cash for them all, but mostly for Ben. It had been to secure his future. He was a child with severely limited speech and behavioural problems. But she knew the way only a mother would, that if he'd been able to understand he would have given Jonathan his blessing too.

'I love you,' she quietly said to herself as she stood a few moments on the doorstep. 'I'd gladly have gone with you. But I have to stay with Ben.' She shrugged her shoulders as she turned to go inside. She sighed, 'I trust you to do what's right

… always. We're with you. God is on our side. I know it. I feel it. It'll be all right. Just wait and see.'

∞

Jonathan arrived with minutes to spare and found himself a seat. As he sat down he heard British Atomic Management chief executive Sir Jack Geisner introducing that night's speaker, the recently knighted Sir John Gill, KBE FRS. He'd once been one of the country's leading scientists, working with the father of modern quantum theory, Paul Dirac and at Princeton, New Jersey, with physics Nobel Laureate Murray Gell-Mann, before abandoning his scientific career to become an Anglican priest, in which capacity he was now simply known as Reverend Doctor John Gill.

He was well into his late seventies and was wearing an Anglican dog collar above his tweed jacket. He had thinning, wispy white hair, but an overall appearance that belied his years. He was as alert as a man half his age and still had a humorous twinkle and fire that shone from his eyes.

He spoke eloquently for an hour, finally concluding by quoting noted English-born physicist and mathematician, Freeman Dyson, when he had said 'Science and religion are two windows that people look through, trying to understand the universe outside. They give different views, but they both look out at the same universe. Both views are one-sided. Neither is complete. Both leave out essential features of the real world. Both are worthy of respect.'

Project head Sir Jack Geisner then came back into the lecture theatre and thanked Reverend Dr John Gill, before asking if anyone had any questions. There was the usual mute silence. It was Dr Jonathan Anderson who stood up and in a clear voice asked:

'Earlier, you touched on the beauty of mathematics in cosmology. Do you think this beauty is mere coincidence?'

Reverend Dr Gill smiled as he recognised the speaker. 'It's a good question. But I think we know one another. Am I right?'

'You were my tutor at Cambridge, Dr Gill.'

'I remember. You were a high-flyer if I remember correctly.' Dr Gill paused a moment while thinking Anderson's question over. He looked down at his feet before walking round to the front of the lectern. He then stroked his chin as if still in deep thought.

'It's a good question,' Dr Gill repeated, before saying: 'No, I don't think it is a coincidence. Mathematical beauty frequently occurs in describing the exterior, physical world. There seems to be congruence between what we experience within our minds and in successful descriptions of the external world. Here, I am talking not about the banalities of everyday life, but about the ability of mathematics to unlock the deepest secrets of the universe, which seems to require both economy and elegance in any successful, mathematical description. This occurs to such an extent that mathematicians call it "non-trivial". In other words, it is significant.'

He looked directly at Anderson as he paused for a moment. 'One of the easiest ways of condemning a colleague's work is to describe it as either "ugly" or "contrived". Successful theories have elegance about them. It is not just sporadic or occasional. The elegance seems to be inherent, a requirement of the explanation. Now why should that be so, given that modern mathematics can be so very strange?

'Something bizarre seems to be happening, something connected with the deep intelligibility of the physical world, far beyond what evolutionary biology can explain. It seems that this physical world is shot through with a mind. And this possibility, speaking as a Christian, is highly satisfying. It means there seems to be an underlying rationality behind what we see and experience. In other words the mind of a Creator.'

It was enough for Jonathan. He had seen the 'beauty' Gill described in the beauty of his own equations. But they had pointed to other things too, things that had not been there before. Collectively, they meant that though BRIGHTSTAR held the promise of rendering nuclear war obsolete, it could be too dangerous to ever use.

There were other questions asked, but Jonathan was no longer interested. He'd got what he wanted. Gill had confirmed where his own thoughts had already led him. It only left him with one further option.

Afterwards a light buffet supper was served giving the speaker an opportunity to meet with senior staff. As Head of Applied Laser Fusion Jonathan was among those invited, even if just out of politeness he wanted to say 'Hello' to Dr Gill. It took a few moments to find him talking with Sir Jack at the far end of the room where the supper was being served. Sir Jack raised his arm in an affected greeting when he saw Jonathan, and immediately introduced Reverend Dr Gill, forgetting that the two already knew one another.

'Hello, Reverend Dr Gill,' said Anderson. 'Good to see you again after all this time.'

Reverend Dr Gill smiled warmly and shook Anderson's outstretched hand. 'Although the audience was polite tonight it's always good to see a friendly face.'

Sir Jack then cut in. 'As you two already know each other, if you'll both excuse me there's someone else I really must speak with.' He nodded vaguely in the direction of a group standing five or six metres away, and without waiting for a response began shuffling in their direction.

'He's a busy man,' said Anderson, as they looked at Sir Jack's receding back.

Gill smiled benignly, then said, 'I didn't know you'd come here after Cambridge. I'd always seen you as staying in academia.' There was a question in his statement.

'It's a long story,' Jonathan replied. 'But I think tomorrow could be my last day, and academia might be the only area left that'll have me.' He paused a moment to let the words sink in.

'Sounds like it could be a question of morality? If it is, then it's brave of you,' guessed Gill.

'It's okay, Reverend,' said Anderson, seeing the slightly concerned look in Gill's face. 'You're right, but I can't say any more. Secrecy and all that. It won't matter after tomorrow. I'm planning to follow in your footsteps by going to theological college next year.

'Well, I'm sure the church will be glad to have someone of your stature. I assume and hope it's Anglican?' Dr Gill allowed a wan smile to cross his face before adding graciously, 'Anything I can do to help?'

'No, thanks. It's kind of you to ask. But in a way you already have with tonight's lecture, and by answering my question.' Anderson looked into his wine glass with a troubled expression. 'I just wanted to know that I'm doing the right thing, and that I was right.'

'God will know,' responded Dr Gill.

'I hope so,' replied Anderson.

'But you might care to remember the fate of Dr J. Robert Oppenheimer. After being midwife to the atomic bomb he later opposed building Teller's hydrogen bomb. It didn't end well for Oppenheimer.'

He took a last sip from a near-empty glass of tonic water before looking at his wristwatch. 'Morality never had much sway in the physics of Oppenheimer's day. I'm nor sure if it does today,' he added.

Jonathan replied, 'I think what you were saying Dr Gill is that God seems to be on the side of the beautiful.'

'I think so,' Gill said as he put his glass down on a nearby table before adding, 'And maybe that should be enough for us. For us physicists at least.'

He touched Anderson on the elbow then added that he really must be going, wished him good luck, and then turned to leave. 'It's a long way home Jonathan and it's my wife that has to do all the driving as I don't drive myself. So it's for her I must go. But first I have to find where she is!' He muttered more apologies and then left to begin his search.

Anderson left shortly afterwards. His hope was that there might still be enough time to talk over things further with Katrina. She often worked late, reading books or other papers.

∞

It was 9.30 when Jonathan cleared the main security gates at Burghfield. It normally took an hour along the winding roads to Ripley. There were few other vehicles. The stars sparkled in a wintry sky, while a pale moon cast its wan light over frost-covered fields. As he drove, he remembered he'd forgotten to switch his mobile 'phone back on after turning it off for the lecture. He stopped the car and fumbled inside his jacket pocket. The screen lit up. There were three messages: the first a call from Katrina; another from his brother Edward; and the last, a follow-up text, again from Katrina. He read it quickly.

> *Ben suffered seizure tonite.*
> *Ambulance taken both of us 2 Atk Morl.*
> *We R both here. Might be 4 nxt 48 hours.*
> *Part of condition. Not to worry.*
> *Speak later.*
> *Love Kxx*

The message was timed 9.05 p.m.

∞

Professor Robert Pahlavi looked down through his bifocals at the two dark grey, flat and circular objects nestling together in his trembling left hand. There was loud music playing somewhere in the bar, making it difficult for him to think. He scrutinised the objects as if he'd seen them for the first time. Each measured roughly forty-five millimetres in diameter and was seventeen millimetres thick, making them the same size as a small roll of Sellotape. In their rounded sides each had what a computer expert would have recognised instantly as abnormally waterproofed, series 3.0, USB computer connection ports. They allowed data transfer rates of 5GB per second; ten times that possible from any similar available high-street device. The discs were almost featureless until he turned them over. On their obverse sides, in small, neatly stamped lettering, was the legend:

Property of United States Government.
Return to Los Alamos National Laboratory.
Shipping address: PO Box 1633, NM 87545, USA.
Serial No: 175890/1603/11.
Possession is an offence without proper authorization.

Professor Pahlavi was nervous as he waited in the bar of "The Spinnaker", on the waterfront of Sausalito, across the Golden Gate Bridge from San Francisco. It was half-crowded, not a bad result for a Monday evening with a Californian economy in the toilet, and a governor trying to stave off state bankruptcy for the umpteenth time.

"The Spinnaker" was where he'd been told to go. He glanced down again at the two discs before replacing them back in his jacket pocket. It wasn't that he was going to give them to anyone, but he might just show them. The lettering he'd just read for the eleventh time, stating who they belonged to, could be the means of establishing the truth of his words: of establishing he was of limited value, a professor of vulcanology, not a person associated with anything military or involving high-technology secrets.

If they wanted him to spy then he had nothing to spy on. There was nothing classified about his work. Certainly not at Los Alamos where, although he knew much of their work was classified, he was not in a position to gain access to those parts of the laboratory site. Not even close. It was useless to even try. All it would do would be to alert security, which would lead to investigations and questioning.

He looked out through "The Spinnaker's" large, plate-glass window, affording views of the Golden Gate Bridge and across the bay to the city. He grabbed hold of his bottle of Bud Light and took a long, slow gulp. Outside the window he could see an early-evening fog swirling in from the open sea, soon to obscure his panoramic view of twinkling lights on the other side of the bay. He thought he could make out the TransAmerica Pyramid, or it could have been the Bank of America on California Street. But as he looked they vanished beneath a cloak of fog.

'Mr Pahlavi?'

'Yes … er … umm …' Professor Pahlavi stammered nervously for a moment, not knowing what to say or who it was that was addressing him. He looked up. A glance was enough, and told him what he needed to know. The man was obviously Iranian like himself. He was young, with dark hair and skin. That was all Robert Pahlavi noticed in the poor lighting of the bar. The man made himself comfortable and placed his own bottle of Bud Light alongside that of Professor Pahlavi's. It was the prearranged signal.

'We have some things to talk about, Mr Pahlavi,' the man said in almost flawless English.

Professor Pahlavi noted that twice the stranger had used his name and each time had omitted his formal title of 'Professor'. It was their way, he thought. Their way of trying to make him seem smaller than he was. No matter, he had always expected a contact to come sooner or later, from someone from his native Iran. He had lived in fear of it, of what it would portend. He'd never known how, when or where it would come, only that one day it would. He had hoped against hope that he would be wrong, even more so since they had moved to America from Britain.

He was not, and never had been, a brave man. He was short, with thinning dark hair combed straight back from his forehead. He wore unfashionable half-framed spectacles and walked with a gait; it would have been a swagger if he had been a larger man. But he was a scientist. He was not and had never wanted to be a soldier; frontline or any other. Science was where his heart lay. It was what he loved.

It had been ten years since he and his family had first fled Iran to London, after the first warning concerning his safety had come from the Dean of Tehran University. It was because of some long-forgotten photograph of himself shaking hands with former ruler, the Shah, the Dean had advised. It had appeared in the English-language daily newspaper, *The Iran Times*. 'You will no longer be safe here,' the dean had said in a manner implying deep, personal sadness.

His father-in-law had made the arrangements at his wife, Shirin's, request. Her father was rich as a result of moving his vast construction-industry fortune out of Iran, well before the mullahs had come to power. He, too, had had gone to live in Baltimore, with his own wife and three sons.

More quickly than Professor Pahlavi had ever dared hope, it had all happened. At the airport no one had asked any awkward questions, there were no police to stop them. Later his wife had said it was just 'money'.

'Enough money buys everyone,' she'd said. 'Everyone has a price.'

He'd been able to resume his old life style as a professor of vulcanology at London's Imperial College, just as he had been at Tehran University. They knew of him at Imperial for he had visited many times, attending symposiums, sometimes as a speaker, and to visit old friends, one of whom had been a colleague in Tehran before, like himself, being forced out.

Imperial College had given him a future again. Gone were fears of a knock on the door in the middle of the night, of disappearing with the secret police never to return, as had happened to some erstwhile colleagues. His only regret was that his eldest son had refused to leave with them, preferring to stick with his new wife, baby son, and plans of becoming a leading surgeon in his own country.

Despite their constant fears for him, they had lived happily in Britain, until the unexplained deaths of one or two dissident Iranians in Paris, who had once opposed the mullahs' rule.

Shortly afterwards, he had received an innocuous-seeming email from someone in Tehran, asking for a meeting; nothing more. It was followed by a visitor he had assumed was connected with British state security, warning him of possible approaches from unknown Iranians. People who might request things from him, or ask him to undertake certain actions prejudicial to 'the safety of the realm'. Professor Pahlavi assumed they had read the email. They, the British, had long experience of such matters, they had said, going back almost a century, with the formation, in 1909, of the Anglo-Persian Oil Company, later to become British Petroleum.

The wily British had given him a mobile telephone and a laptop. Inside the laptop was a special electronic chip that stored and tracked all emails sent to him; however, the sender tried to disguise their origins. Additionally, the chip would encrypt all data stored on the PC's internal hard disc so that no one, 'and I do mean no one', a 'Mr Jones', the man from their secret services department, had emphasised, 'will ever read it … unless we want them to'. He'd paused to let the information sink in before continuing with the rejoinder, 'Use it at all times. You never know when a message will come. It might look quite innocuous at first, but one day it might be all we have to save your life, or even that of your son in Tehran. They'll use him to get to you. It's how they work. We have decades of experience.'

'But why would they want me?' he had asked.

'Because of your specialist knowledge of your country's geology, of its earthquakes, which will still be useful to them. Or it might be simple, old-fashioned espionage, using your son's life as leverage.'

Fear had been drilled into him from that moment on so that, even now, he still used the same laptop, updated continuously, because he knew everything on it would be secure. Just as well, he had sometimes thought, because no one had ever checked it, even at the Los Alamos security gate. Their focus was always on the two discs: when he checked them in and when he checked them out.

He could understand why. It was the actual data they wanted secure, not the laptop. Without the two together and two further encryption systems, one at Los Alamos and the other specially installed at Berkeley, the discs and laptop were useless. They all needed to connect together to work. Professor Pahlavi had sometimes wondered whether the Americans had removed the British chip and replaced it with their own, or whether there were now two quite distinct levels of encryption: one American and the other British. All he knew for sure was that his PC still seemed to work fine.

Shortly after the mysterious email and the British security official, the opportunity had arisen to set up a new department at Berkeley, California, and he'd jumped at it. He'd gone ahead of his wife. She and their twelve-year-old son had followed four months later, after selling their London property and arranging for the furniture to be transported.

Since then, he'd heard nothing from people he did not know. He had hoped it would stay that way. It had ... until a week ago.

The contact had come not through an email but through what he thought was a chance encounter while on his way from his home to the computer warehouse MacAdam, one of San Francisco's largest home computer dealerships, sited on downtown Folsom Street, where he had gone to collect a new graphics package. A man had been begging outside for money to help refugees from Iran; people like himself, but poor. He'd stopped and handed over twenty dollars. The man had given him a folded news-sheet describing his group's activities.

When he'd got home, inside had been a hand-scribbled note giving a mobile phone number to ring urgently if he valued his son's life in Tehran.

He did not know how they had located him, nor did he care. He remembered a wave of nausea sweeping over him. He had not cared how they knew. All that mattered was they knew. He did not care either how his son was wrapped up in it, only that he was alive.

He'd told no one. All he knew was the message spelled trouble: the fewer people who knew about it the better. The fewer people to explain to, the fewer people one needed to trust.

The newcomer and the professor talked nervously about everything and nothing. How life was in Tehran, the progress being made, how the ultimate rulers, the mullahs, saw things, and how they hoped people like the professor would still help their country. They talked for thirty minutes whilst all the while the professor little realised he was being lured into a trap. Meanwhile The Spinnaker had filled with more customers making it progressively more difficult to talk above the din.

Finally the newcomer stood up. 'It has been a most pleasant evening, Mr Pahlavi, and on the basis of what we have discussed I can see no reason for us to fall out.' A faint smile flickered across his face as he spoke. 'But alas, now I must go. There is much to do. We will meet again, soon. My friends in Tehran, I know, will be pleased. Your son will be safe. So, all of us need not worry unduly.' He stood up and extended a hand to Professor Pahlavi. It was not accepted, so the man withdrew it quickly. 'Have it your own way, Professor ... until we meet again.' He turned and walked a few yards before being swallowed up by the crowd.

∞

They always worked as a team of two. It was the way of the SAVAK, as the Iranian secret police were colloquially known by some of its operatives. It was ordained in their new procedures; not as he would have wanted, but that was the way it had to be. Those were the orders. So that was the way they worked. Farhad was the 'contact' and he, Kouros, was the 'sweeper', keeping a long distance behind, ready to deal with any followers Farhad might have been unfortunate enough to have attracted.

Kouros turned the whole matter over in his mind as he sat, feet on the table in front of him, waiting for Farhad's return, ready to tell him what he had been forced to do.

Farhad would not know what had happened to this particular follower, who would not now be following anyone any more. He had been despatched to meet with Allah. What did it matter? What did it matter how many he shot? They were all part of the Great Satan's army, as their chiefs called them; the country, America, that had done so much to inhibit their plans to rise again and be the nation they had once been in their glorious history. The Great Satan were those who plotted and schemed against them, along with the satellite nation of the Israelis; the nation that would do anything to keep American money flowing, to keep them alive in an otherwise hostile, Arab world.

Acting together, often with the British, even now they were preventing the development of Iran's nuclear programme; had even turned the Internet against them with the release of their Stuxnet virus against their Bushehr nuclear plant's process control computers. It had unfortunately wrecked countless gas centrifuges vital for enriching uranium for their atomic bomb programme. They were at war already, in all but name.

Because of it, Kouros had no conscience about what it was he must sometimes do. Tonight had been one such occasion. An American was now dead.

He had followed him in his car, allowed him to get in between himself and Farhad. He'd hung back, watching him all the way over the Golden Gate and into Marin County, still a few cars behind at the bridge tollbooth, then along the winding road into Sausalito.

He had watched his quarry position his car adjacent to the Bank of America branch opposite the private road on which The Spinnaker was situated, had observed him until there could be no mistake he was FBI or similar. He was a soldier like himself, ready to die if the need arose. What did it matter? His quarry was young, Kouros knew that. Maybe inexperienced too? Kouros had got out of his own car. Most of the square was in shadow, for the lighting was poor. He had skirted round behind him, and knocked on the driver's side car window. The man had lowered it, smiling.

'Yeah, bub?' was all he said. They were his last words.

He had shot him with his silenced 9 millimetre Glock 17 hand pistol, in the left temple, he believed; a small hole. The man had looked surprised, the expression frozen on his face as he had slumped forward over the steering wheel. To cover the blood of the exit wound he had reached in and placed the hat off his own head over the head of the man. It was a nice touch, he thought. There had not been much of a sound, no witnesses ... nothing. Textbook example, really.

The driver's-side window had remained intact. He'd shut it afterwards, as the agent's body lay slumped over the wheel looking like he could've drunk too much. He'd taken his wallet too, to make it look like a robbery. Maybe drugs-related, the police might think. Overall he'd been quick and he'd been efficient. Kouros felt pleased.

∞

Maybe it was because FBI Special Agent Ryan Martinez Jnr was bored. He'd driven out to Berkeley, he'd driven back, he'd waited and he'd watched the professor all day, just as he had done for many months past. Today was only slightly different from the routine.

After driving back from the Berkeley University campus the professor had gone out on his own; not to shop for computer-related gadgets this time, but over the Golden Gate to Sausalito, to a nice bar and diner. Martinez stopped on the opposite side of the square, partially hidden by the clump of trees at the square's centre. He could still see the professor's car through his own windscreen. Must be nice to afford The Spinnaker; its windows gave the best view across the whole bay to the city. He'd watched the professor go in, seen him joined by someone he'd never seen before. He'd scribbled a quick note of it.

Maybe it was the months of long babysitting he'd already undertaken on Professor Pahlavi. Maybe it was the dull, repetitive lifestyle of the professor, interspersed with the occasional flights to Baltimore where the professor met with his wife's family. Often she went on her own while he stayed behind, working. He knew it was work through shadowing him to Berkeley from their home in expensive Bernal Heights, part of downtown San Francisco, within easy reach of Bay Bridge and the quick access it gave to the Berkeley campus. The FBI profile said her rich family had paid for the apartment. How he'd sometimes wished for a rich dad of his own rather than the patrolman father he'd got. No, that was wrong. He loved his dad, but wishing was sometimes nice.

Another relief from the tedium of the professor's routine was when he shadowed him to New Mexico, all the way to the Arrivals lounge of Albuquerque's Sunport airport. He'd usually lurk well behind until he saw Special Agent Tom Santos take over.

Sometimes it was just overnight, and then back to San Francisco in the morning. He stayed at a local hotel, a Howard Johnson usually. They were cheap. Sometimes he flew straight back and then waited for the professor's return with Tom Santos always in evidence, his battered Stetson hat making him stand out. There was never anything clandestine or furtive about him: cowboy boots, check shirt and all.

They never approached one another, never acknowledged each other's existence. All he knew was that the agent's name was Tom Santos, with a clearly different relationship with the professor than his own. At Albuquerque, after the flight from Oakland touched down, he'd watch Santos walk over and shake the professor's hand. What was that all about, he'd wondered. He'd been shown photographs of Santos for recognition purposes, and knew he'd once been with a special CIA black ops unit before he'd joined the National Intelligence Directorate. But, strangely, he'd retained his position as a US federal marshal. Very strange.

Special Agent Ryan Martinez Jnr had been waiting forty-five minutes for the professor when he decided to write up his notes for the day, like he did every day. Today he was later than usual, not that there was much to write.

The square was quiet. Few people were about. It was dark and cold even for San Francisco. He remembered what American writer Mark Twain had said: 'The coldest winter he'd ever spent was a summer in San Francisco!' Wasn't that the truth! It had been a long and boring day. He was tired and pissed off with its uneventfulness. Suddenly, and from out of nowhere, came a sharp knock on his driver's-side window. He looked up. It was a foreign-looking guy. Now, what the fuck did he want?

It was all over in seconds.

CHAPTER 5

It was two days later. Present in Sir Jack Geisner's office, in the main Aldermaston office complex, were Sir Jack, the company's legal advisor, and the Head of Security. They were nine minutes into the meeting, and sat comfortably in leather armchairs around a low glass-topped table on which a tray of now half-empty coffee cups and biscuits had been served five minutes before. It was Sir Jack's first meeting of the day. He'd been in his office since 6.00 a.m.

'We've got to have legal redress against this bugger,' Sir Jack was saying as he replaced his empty coffee cup on the tray. 'We've got an employment contract with him, for God's sake. Surely that's binding on him as much as it is on us?' His voice sounded more American than Australian.

'It's a good thought, Sir Jack, and in theory it's sound. Problems start when you try to enforce it,' replied the company legal advisor; a tall, thin man in his mid-thirties. 'I've got the contract here.' He lifted the buff-coloured file off his knee. 'I've been through it. Standard stuff, as I would expect. Three months either way. Anything more onerous and you'd be up against European Union human rights stuff. You don't want to go there, take it from me. As for the Official Secrets Act document,' – he sighed – 'well, what can you do with that? If he honours it then it'll give you no extra leverage. And if he doesn't, then you'd first have to prove it. No easy task unless you caught him red-handed, on camera. I think that would be highly unlikely in my experience. And anyway, at the end of the day, what's the point of trying to force someone to work for you when they plainly want to leave?'

'In the States he wouldn't get away with it, I can tell you,' Sir Jack snorted in derision. 'He'd be banged up good and tight in the state penitentiary. Questions would be asked later, or maybe not at all.'

'Well, that's what the cowboys might do in the former colonies, but not here in civilised Europe.'

There was a temporary lull in conversation. So far the Head of Security had not uttered a single word. Now it was her turn.

'Out of curiosity, Sir Jack, what reasons did Anderson give for resigning?'

'Does it matter?'

'It might.'

'Said BRIGHTSTAR was now against his beliefs and teachings and his bloody morality as a Christian.'

'Did he say why he hadn't said that before?'

'Some bullshit about he didn't know then what he knows now.'

The telephone on Sir Jack's desk, at the other side of the room, buzzed momentarily, then buzzed again, stopped, then buzzed some more.

'Shit!' said Sir Jack. 'I told Dilys no interruptions unless it was important.' In ill-humour he jumped up from his chair and strode over to deal with the incessant buzzing.

'Hello. What?' he barked. 'How the hell do I know?' Pause. 'Stall him until I'm finished here.'

He slammed the phone down, before rejoining the other two.

'Sorry about that. It's the minister, the new minister, trying to fix up a tour of the new Morwenstow facility. I think he wants to bask in what he sees as possible reflected glory stemming from our success and efforts.'

The company legal advisor continued where he'd left off. 'I assume you've looked into his assertions that BRIGHTSTAR might now be dangerous?'

'Sure. On a multi-billion pound project, and contrary to what you might read, we don't fly these things by the seat of our pants. We check them, and then we recheck 'em again.'

The lawyer thought for a few moments before responding. 'Nothing?'

'Nothing,' Sir Jack replied. 'If you ask me he's been lured away by a rival team. It's about money. It's always about money, whatever they say.'

The legal advisor smiled a sardonic smile and said quietly, 'It happens in the football business all the time, contracts or no contracts.'

'Wouldn't if I was in charge,' snorted back Sir Jack.

The legal advisor ignored his comment. 'Getting down to basics, do you still need him? Now, or in the future? How much will his departure affect your plans? Is he irreplaceable?'

'Graveyards are full of indispensable people,' quipped Sir Jack. 'Nope, we can get by without him, like you can get by flying a jumbo jet without the pilot … until you want to land, or the unexpected happens, like you hit turbulence. It's like Einstein. Although he was never employed directly on the Manhattan Project the great man would sure as hell have been whisked in for explanations if the atom bomb had turned out to be a dud. And after all the money they'd spent.'

'So there's no immediate worry?' enquired the legal advisor.

'It's not quite as clear-cut as that,' the Head of Security said. 'Hasn't Anderson being trying to piss on our parade for some time, Sir Jack?'

'I suppose you could call it that,' Sir Jack replied. 'They all do when a project like this is nearing completion. These liberal scientists get cold feet, talk about morality. They're like artists: big egos allied with negative social charisma, and a complete lack of guts. Just when we're ready to kick the ball they want to move the goalposts, or take them down completely. We've checked all his claims. They're rubbish. We're this close to roll-out.' He indicated it with his thumb and forefinger. 'That's the way it's gonna stay ... Dr Anderson or no Dr Anderson.'

'Fine, Sir Jack, that's your prerogative. Mine's security; national security, defence of the realm,' said the Head of Security. She was a mean, sour-faced, humourless woman of forty-five with a doctorate in Mesopotamian art: the Akkadian period from 2334 to 2154 BC. Somehow that had qualified her for Britain's security services, finally ending at GCHQ before being seconded to British Nuclear Weapons Management Plc.

Her voice was as cool and calm as menthol, completely lacking emotion, with a hint of cut glass. 'However we cut the cake he'll remain a security risk. And as Sir Jack says, he might be going to a rival team, one who'll pick up all his out-of-pocket severance costs, including legal, if there are any.'

'Where might he be going, if anywhere?' asked the legal advisor, before supplying his own answer. 'The Americans?'

Sir Jack picked the question up. 'It might be worse than the Americans.' He paused. 'It could be the damned French!'

'I thought they were our partners.'

Sir Jack looked up at the ceiling as if seeking divine inspiration.

The Head of Security ignored his actions and continued. 'There's a way to keep him occupied. We don't use Hollywood style heavies any more. We don't need to break any bones these days. Instead, we use the law. What we accuse him of may not need to be true, as long as it's defensible. In the UK there are many laws where you can tie an individual up in knots, freeze all his or her assets for years, and then never charge them with anything.

'Even get the boys in blue in on the act. Get them to do a dawn raid, allegedly searching for hidden cash, documents, anything. Get them digging up the garden, searching bedrooms. It's unnerving for those being searched, especially early in the morning with guys stamping over the flowerbeds and breaking down doors. We just need a sympathetic judge to sign the right papers. That's how we pressure him. He's got a sick son, too; that's his and his wife's Achilles heel. We freeze their assets with a court order and see how long they continue to play footloose and fancy-free.'

The legal advisor looked uncomfortable, and said so. Sir Jack's face, though, had brightened. 'You think we could do it?'

'I know we can,' the Head of Security replied promptly. There was a sound of certainty in her voice, of a type that implied this was not new ground she was covering.

'I'm reliant on you here. You know the Pommie and EC law better than I do,' said Sir Jack.

'If you agree,' she replied, 'Perhaps I can go away and work it up into something and come back later today? Depends on how urgent you see it.'

'At the moment it's bloody urgent. What time? 4.30 suit you?'

'Whatever's in my diary, I'll clear it so we can get on with this,' she replied.

'Do you want me to attend?' said a still uncomfortable-looking legal advisor.

'Nope,' said Sir Jack immediately, in a tone that brooked no argument. 'Let her do her stuff. We'll talk it over between us, then get you involved and tell you what we want doing, if anything. We're just taking precautions. It's an option if we need to keep Mr Anderson in order. No good shutting the stable door if he's bolted, talking to all and sundry with him in the process. As I say, it's just a precaution, that's all. Nothing for you to get worried about.'

There was silence. Sir Jack got up from one of the armchairs and said: 'Good, then. You and I will meet later.'

He looked at his Head of Security. The two rose and walked towards the door. Sir Jack turned and walked back to his desk.

Outside it was still raining, just as it had been throughout the previous night. So far, the year had been unseasonably warm, with a long, hot summer. Sir Jack moved from his desk over to the window to absent-mindedly watch the rainfall on the wide, concrete road, once an RAF runway, but now called AVENUE 4163 EAST.

He had been chief executive for over four years. BAWM Plc was a specially formed consortium of British Atomic Fuels, the American aerospace giant the Dellorto Corporation of Ohio, and another US company, Qume Scientific. BAWM Plc had won the management contract in 2009, from Percival Baliol Plc, after Percival Baliol had suffered two prosecutions for major safety breaches in 2010: one for discharging radioactive tritium heavy water into a nearby stream; and the other when two workers had inhaled plutonium.

The British complex was modelled closely on the American-developed government-owned contractor-operated management, or GOCOM, system of government ownership, using private management. It had been in place for years at similar US facilities, such as the prestigious Los Alamos National Laboratory in New Mexico, or at the 10,000-acre nuclear weapons assembly/disassembly plant at Pantex, Amarillo, Texas.

BAWM's responsibility was for the day-to-day running of the two Berkshire sites, with the added responsibility of Morwenstow and Glastonbury. They had a collective annual budget approaching £730 million. Secret Treasury estimates said it could reach £1.5 billion when BRIGHTSTAR became fully operational in a mere three months' time.

Two thick piles of coloured folders on Sir Jack's desk contained the daily digests, covering the whole BAWM operation of which BRIGHTSTAR, usually referred to as 'the Project', was a big part.

The BRIGHTSTAR folders were pink, and always caught his attention first. They were where the urgency and political pressure lay. The other buff-coloured reports were more straightforward, involving processes necessary for the continuous upgrading of the nation's nuclear weapons stockpile. Mostly this would affect the Trident D5 missiles carried by the four 'Extended Life' Royal Navy Vanguard-class submarines. Each submarine carried sixteen missiles armed with four multiple, independently targeted warheads.

As Sir Jack stared out of his office window he saw an articulated lorry reversing along what was left of the former wartime RAF runway, into the A45 building, where they had once hand-assembled nuclear warheads for the Navy's Poseidon and, later, Trident submarine programmes.

He turned from the view, as if he'd suddenly made up his mind about something, and walked towards his desk, on which two buff-coloured files lay partially opened. Although he began going through the files his mind was elsewhere, on the issue of Anderson.

With him he had always known there was a fall-back position in Professor of Theoretical Physics at Trinity College Cambridge, Sir Robert Carr; a Nobel Prize Laureate who knew his stuff. The intellectual grunt work was long finished. Anderson had already seen to that ... just. What was left was, primarily, assembly and testing, particularly with the line-of-sight link from Morwenstow to the experimental Government reflector station near Glastonbury a hundred miles to the east. Following Anderson's departure he'd immediately accelerated BRIGHTSTAR's development plan. It was already months ahead of schedule. It was going to be spun up to full power over the next three months. It would take maybe two or three trials. Then would come the experimental test against a Royal Navy destroyer somewhere out on an MoD missile and bombing range in the Irish Channel. So far it all looked on track and on schedule. After that it could all be handed over to the new Department of Homeland Security tasked to be the body that would look after it on a day-to-day basis.

Sir Jack knew that among his many other responsibilities, all of which he likened to keeping plates spinning, there was now another 'plate' in the form of Anderson. Sir Jack wasn't complaining. Tackling and solving tough issues was what he was paid for. That, and getting BRIGHTSTAR operational on time and on budget. His contractual obligations and large bonus payment depended on it.

Anderson had formulated a mathematical insight into how pulsed lasers could be used to achieve fusion in a careful and controlled way. Anderson had said it potentially made BRIGHTSTAR the 'laser cannon' featured in films such as Star Wars or Star Trek. It potentially even gave verisimilitude to the US's so-called, and mostly fictitious, Star Wars programme of the late 1980s, launched by the fortieth president of the United States, Ronald Reagan.

At their last and as it turned out final meeting, Anderson had gone through much of his mathematical reasoning in a quiet and controlled manner. Most of it had been way over Sir Jack's head. He was there to manage, everyone else was there to understand ... but he understood enough.

'The projectors, the laser cannon, might be leaky,' Anderson had said. He'd never appreciated the possibility before because it was only developments in new mathematical techniques, more familiar in cosmology, that now allowed such possibilities to be envisaged. Anderson had emphasised it was 'only a possibility', no greater than that. But even so, in his opinion, the risk was unacceptable.

He'd concluded by saying that these new techniques, involving quantum theory, indicated that no longer was there a single universe. It had been replaced by a 'multiverse' where all times coexisted alongside each other, where all possibilities occurred, where all histories happened together. His calculations showed that the intensity and power of his laser cannon might not only be capable of downing incoming ballistic missiles in flight, as hoped, but also of punching a hole from one history into another. They would have no way of knowing until they tested it; maybe not even then. It might be that unpredictable, as his extended theory now suggested.

To add further complexity, almost as an afterthought, he'd added that the human brain functioned ultimately as a bio-electrical quantum computer. It might well be the mechanism where ideas and dreams originated. What would happen if people were accidentally 'illuminated' in a beam's or cannon's path? As they might, given the paths some of the beams might take. The possibilities were remote, but still a risk.

Sir Jack considered himself an expert on the Manhattan Project, so he knew the historical precedent to Anderson's decision. Prior to accepting the BAWM role he had, for eight years, been director of the clean-up operation at the American Y12 gaseous diffusion site at Oakridge, Tennessee.

It had been bad there. The American public only knew the half of it. Thirty years of enriching uranium for transport to Hanford in Washington State, to make plutonium for nuclear weapons, was always going to leave its mark, and it had. It would be there for another half a million years.

He remembered reading how Los Alamos scientists had reacted as the Trinity Test drew close. Like Anderson, some had begun having second thoughts about its actual use. It was an interesting precedent, he thought.

Meanwhile, in the race to harness nuclear fusion, the Americans had chosen a different route and focused on their Lawrence Livermore National Laboratory, in California, where the name of the game was 'inertial confinement fusion' and the whole facility had been called the National Ignition Facility. The site covered an area several football pitches in size, with, so far, little to show for their billion-dollar investment.

Sir Jack Geisner was physically a big man, standing six foot four inches in his socks. But looks could deceive. Thirty-five years before, he had obtained a double first from Cambridge in physics and mathematics, before setting out to make an indelible mark on the world of big engineering projects.

Sir Jack had reflected briefly on the possibility of having Anderson's calculations rechecked again. Nothing had been found the first time. He'd concluded that it was too late to stop the project and, worse, might let the BRIGHTSTAR cat out of the bag. When the Americans found out, they would want to muscle in. Last time they had thrown their weight around it had been with the Manhattan Project, which had only succeeded after prodigious British input.

After two A-bombs had been dropped, the ungrateful Yanks had kicked the Brits out by passing the 1946 McMahon Act. It had taken ten years to repeal, and that had only been when the Yanks wanted British-perfected ways of miniaturising American, ox-cart-sized H-bombs, to fit inside conventional missile or torpedo warheads. The British had been bitten once. They wouldn't risk a second time.

Sir Jack suddenly remembered there was still the issue of the ministerial visit he had to fix, which his PA had mentioned earlier. He reached for the telephone.

It was 10.19 a.m. when he opened the first of the pink folders. They would all be cleared within the hour.

∞

Forty-eight hours later, the door of the small hospital room opened and in walked Jonathan Anderson. Most of the space was taken by a now-empty patient bed, together with a foldaway Z-bed that Katrina had been using until that morning in order to be near their son; there to comfort him if comfort was needed, to be there just in case, to soothe him in his usual night terrors. She had done it at many hospitals, many times before in his short life. She had grown accustomed to it, as had Jonathan.

The room was bright and cheerful, covered with children's paintings, the floor strewn with toys and a bright colour scheme reflecting care and fun. It was Jonathan's second visit during their son's latest 'episode', as the doctors called them. His first visit had been two days before when he had driven over immediately, following the vital Aldermaston lecture.

Katrina had her coat on and was sitting expectantly in a chair, waiting for him. In a wheelchair was their son Ben, his face alive and alert, his head nodding first one way, then the other; his limbs in seemingly constant turmoil. His body looked broken and hunched, like an outsize but deranged rag doll. Yet for those who knew him, he was not broken. Every new day was a gift he relished. He did not need sorrow but perhaps instead compassion and understanding for the restrictions he lived under. Even more, he needed whatever anyone could give so that he could enjoy every minute, every hour, every day of his life. Somehow he radiated his zest for life. And somehow too it suffused the air and the spirits of those around him. Long ago, his parents realised this gift and rejoiced each day he lived, and were grateful they could share it with him. Yet they knew, too, the daily pain he lived with, which permeated his constantly moving, rogue muscles.

When Ben saw Jonathan he let out a series of joyful and excited yelps and cries, as if he had not seen him for years. His body became animated against a background of continual movement. Smiling, Jonathan moved forward with his arms outstretched to hug his son as if they had been apart for years. Katrina watched, tears at the corners of her eyes, then running down her cheeks uncontrollably.

They were together. It was all she could think of. As Jonathan grabbed the handles of his son's wheelchair and began wheeling him towards the door there were few happier people in the world at that moment. After picking up her overnight bag Katrina opened the ward door, allowing Ben and Jonathan to pass through. Barely had she done so when the senior consultant responsible for monitoring Ben's condition for the previous two days, as well as the past three years, appeared clutching a sheaf of papers together with a thick, buff-coloured file. He motioned Katrina back into the room. She shouted after Jonathan's diminishing figure as he headed for the outside.

'I'll catch you up in a minute. Wait for me in the car.' Jonathan waved his arm in acknowledgement without turning round.

'I was hoping to catch you before you went,' Dr Orum said gently. 'Normally, I don't have much to say after these events. We both know that it's part of his condition. We've done the usual tests.' He nodded down at the papers he was carrying. 'There's little change in his condition, which is good and bad news: good that he's no worse than he was before; bad, because it's exactly to be expected.'

Katrina listened, hoping intently, as she always did, that the news might, miraculously, be better; that there was some sign of an improvement or even a cure. But she also knew it could never be. Parts of Ben's brain had been destroyed by a lack of oxygen at birth and would never grow back. He would be as he was now for life, and that life could be long. She'd been hearing similar statements throughout Ben's life. As a result, she and Jonathan knew as much about his condition as anyone: how he had suffered from oxygen starvation during birth, through negligence in his delivery; how they had, so far, spent six years trying to win compensation from the hospital involved, without any luck to date, except for an increasingly thick file of lawyers' letters and fee notes.

It was Ben who was paying the price, who would always pay the price of the hospital's negligence, by being afflicted with incurable cerebral palsy.

'Thanks, doctor. It's what we both expected.' She moved towards the door ready to join her son and husband. The consultant motioned her back. 'I'm sorry. I didn't make myself clear. I said there was little change, not that there had been no change at all. We've noticed a shadow on his latest brain scan. It may be nothing. It may be something. That's all we can say at the moment. I just wanted to alert you in case it might be life-threatening.' Dr Orum didn't stop there. 'I wanted to add that we have a new technique. Diagnostics, I'm afraid. No cure or anything like that. But our new tool might tell us more about Ben. Knowledge gives us hope. I'd like you to bring him back to us in a week or two's time, if that's okay? We should have our new equipment installed by then.'

'Sure,' Katrina heard herself saying. She'd heard similar tales before and they had all lived through them. She would tell her husband and they would live with it until it materialised into more ... or less. What else could they do?

Katrina had been born an incurable optimist, just as she had been born with a good brain, good enough to have found her way to Fitzwilliam College, Cambridge, and gain a Classics degree, involving the study of the ancient worlds of Greece and Rome. While there, she had met Jonathan and they had formed an immediate bond of intellect, humanity and humour. She was quick-witted with a good turn of phrase. Beneath the elegant exterior she was strong-willed, tough and caring. She did not give up easily. Neither did Jonathan. It was one of the many things she liked about him. For Jonathan, it had been her strangely coloured violet eyes, the same colour as the late Hollywood film star Elizabeth Taylor. For him they had made her irresistible.

She looked the consultant in the eye and wondered whether there was any point in asking him further questions to which she already knew the answers. She decided she'd spent long enough at the Atkinson Morley and decided against it, thanked him for the information, and left to catch up with Jonathan.

She and Jonathan already knew more than enough about the four classifications of cerebral palsy. Ben's was the more common spastic variety, usually causing impaired muscle movement of the lower limbs. In his case, it affected his upper body as well, making him a quadriplegic. Additionally, and not uncommon for his condition, he had developed severe tonic-clonic seizures over the past eighteen months. So far he had suffered five such 'events'. Over his life he could expect hundreds more.

Fortunately, and because of the previous four occurrences, Katrina now knew how to deal with them and which medicines to administer. Importantly, she had been mentally and physically prepared for the sudden onset of crying out and shouting, followed by loss of consciousness, muscle twitching, violent shaking and convulsions. As on the previous four occasions, he had recovered quickly, leaving him confused and, fortunately Katrina now thought, with no memory of what had occurred.

Both Katrina and Jonathan accepted that at the Atkinson Morley Neurological Unit Ben was in the most capable medical hands in the country. Its pre-eminence in its field had grown so that it was necessary to have its own helicopter landing site to take emergency neurological cases from all over the country. More recently, its facilities had been expanded with the addition of the Wolfson Neuro-Rehabilitation Centre on the same site.

Katrina Anderson rushed down the long corridor, down the steps and out into the car park to be with her family, unaware of what was waiting for them. All she knew was that like her son, it was good to be alive.

CHAPTER 6

Dexter Coleman's chauffer-driven black limousine pulled out from Georgetown's Wisconsin Avenue. It headed on to Whitehurst Freeway with the intention of driving along Pennsylvania Avenue from where it joined Washington Circle. The traffic, even at that time of morning, was always an issue. If it was going to be like any of the other times he'd made the journey it was going to take thirty minutes, plenty of time to make his 6.30 a.m. meeting with the president at the White House. The air in Washington DC was cold and frosty.

Coleman was the recently appointed director of the US National Intelligence Directorate, the fourth in as many years and at forty-nine its youngest. He yawned, then picked up that morning's copy of *The Washington Post* lying next to him on the limousine's rear seat. He'd received a telephone call the previous evening from presidential Chief of Staff, Cliff Wurzburg. He'd told him about a story that was going to run the following morning.

'*The Post's* White House staffer gave us the head-up earlier this evening. Gonna be on page seventeen, he told us, complete with photographs. Be sure to have read it by the time we have our meeting tomorrow morning.' Then the phone had gone dead. It was Wurzburg's way. Short. To the point. No small talk.

His car was now into heavy traffic. As he'd thought, there was going to be plenty of time to read the article. He picked up his copy of *The Post* and began scanning its pages. What he was looking for was in the business section. He quickly found it and noted the feature had almost a page to itself.

WHO'S STOLEN THE STATE'S SALMON ?
Staff Reporter

In the Pacific Northwest something wicked this way comes. But among the multiplicity of federal and state environmental agencies involved not a single clear idea has so far emerged as to its identity or cause. And if they do know, then they're keeping mighty quiet about it. Meanwhile one thing is certain, the region's once booming salmon stocks are 'mysteriously' on the decline.

Ever since 1972 at Seattle's Ballard Locks they've been counting the salmon jumping up the ladder as they returned from the open ocean to Lake Washington. Although seasonal fluctuations are to be expected, overall the trend has been forever downward, so much so that the once plentiful Sockeye is now being labelled as an 'endangered species' in some quarters. The important question is why?

200 miles East of Seattle is a 51 mile section of the Columbia River, the nation's third biggest, stretching downstream from Priest Rapids Dam to Richland and which, due to chance location of a World War Two nuclear bomb making facility, had to be left untouched. The result was coincidentally good for salmon although the plant's rationale was that the manufacturing processes involved needed a plentiful supply of cool, clear and uncontaminated fresh water.

The site was called Hanford. This was where they made highly toxic nuclear weapons grade plutonium. During its near 40 years of continuous production they made 64 metric tones of it, enough for the nation's once mountainous 60,000 stockpile of nuclear weapons, now much reduced thanks to the Cold War's end.

The unintended result of the original bomb making decision regarding the river was to underpin the whole region's economic health through boosting salmon catches. This so much so that according to Washington States's Department of Ecology, the Columbia River and its salmon supported 25,000 companies across 9 counties in such cities as Richland, Pasco, Kennewick and Wallula. Out of the 500,000 people living along its shores the river directly supported 280,000 jobs collectively generating a payroll of $9.5 billion. Little wonder then that a member of the state's legislature recently commented in an annual report that "Salmon are as much part of Washington State as our mountains, forests and fields."

Local Native Americans from the Yakama tribe who've lived and fished along Hanford Reach for the past 10,000 years might also applaud as stocks improved, as they traditionally fished at such places as Tah-Koot or Wy-Yow-Na, both close to present day Locke Island sited in the middle of the Columbia. Over eons they provided the indiginous Indian population with seasonal camping grounds from which to catch sockeye salmon.

Whole families would arrive during the fishing season, often as many as 500 people according to one octogenarian. The catch would average as many as 300 fish either for eating fresh as they were caught, for salting and storing for later use when winter came, or for trading with other tribes. Conservation was always an issue with care taken to ensure enough salmon were left to sustain stocks for the following year.

But unknown to everyone at the time, a bomb was ticking. Although salmon stocks were to be an immediate beneficiary of the wartime Hanford bomb making decision, there were to be unintended and sinister environmental consequences. These would become apparent long after the primary goal of obliterating the Japanese city of Nagasaki and its 80,000 inhabitants had been achieved, and for which Hanford had been responsible for producing its plutonium core.

Like most industrial processes, manufacturing plutonium also produces highly toxic waste. But this waste retains its toxicity for hundreds of thousands of years. At the time the then Atomic Energy Commission, aka today's Department of Energy, was none too careful about what it did with it, and the ecology movement had yet to be established. Cloaked in secrecy where questions were not invited, it had carte blanche to do what it liked. The environment was never high on its agenda. Sooner or later the chickens were bound to come home to roost.

Today those 'chickens' are stored in 149 leak-prone tanks containing 53 million gallons of highly radioactive waste making the area the most polluted in North America. And while the Feds and State officials continue arguing over why salmon are disappearing, one thing they are all united over is that sooner or later waste from those leaky tanks is going to find its way into the river. Some believe it already has.

That's why 3,000 people are beavering away on the largest 'design and build' civil engineering project in the country. They are building something called a 'vitrification plant' and watched over by prime contractor to the Department of Energy, Bechtel. One plant won't be enough, so they're building two. Their aim is to turn all that nasty toxic waste into less toxic 'glass logs'. These can then be safely transported to an as yet unnamed underground storage facility to moulder away over the next hundred thousand years. If it's ever allowed.

No state is keen on having radioactive waste stored in its backyard. Nevada once seemed to be one taker bearing in mind it was home to over 1000 underground and atmospheric nuclear tests. Then it seemed to get 'cold feet' so spelling the end for its long planned Yucca Mountain Waste repository. So at present there is nowhere to store the high level, radioactive waste 'glass logs' Hanford will be producing. But everyone at the Department of Energy is fairly relaxed about the problem. After all they have 100,000 years to solve it and which is the half life of the glass logs their plant will be producing. With this length of time scale it gives everyone until the next Ice Age to make up their minds.

The Hanford project is massive covering 65 acres beside the Columbia and is dotted with buildings, some much larger than a football pitch and standing twelve stories high. 65 acres may sound large but is only a small segment of the whole 600 square mile highly contaminated site. The Department of Energy funded by the taxpayer, that means you and me, is picking up the currently expected $12.3 billion bill.

So far the project is more than 10 years late, after beginning life in 1992 with a budget of $1.6 billion.

Latest completion estimates stretch out to somewhere in the early 2020s.

Reasons for this lengthy gestation period are not hard to find. Everyone from the government's Accountability Office to the Department of Energy and on to Washington States's Department of Ecology argue over whether the plant's technology will be safe, adequate and foolproof. Others point fingers at alleged construction shortcomings that may prove dangerous in the longer term. Writs fly, voices are raised and lawyers ensure the courts are filled with 'experts' either defending or attacking every decision.

Meanwhile the main contractor tries quietly to get on with a job that is admittedly technically arduous. And if the plant ever does start operation, then it's going to be running for a long time, some believe for the next 30 years, before its job is done. Perhaps one crumb of comfort is that the British have been successfully operating a similar plant since the early 1990s at their Sellafield, Cumbria, site. Their participation was once sought. But later and for undisclosed reasons they were kicked out.

But time may already be running out. Fish are dying or not breeding as they should because environmental stress – in other words 'pollution' – is causing them to change sex from male to female in their millions, and in what has traditionally been one of their most prolific breeding grounds. Some argue it's all down to higher water temperatures. Others predictably argue that it's not.

But a highly technical and independent report completed as long ago as late 1991, 'Hanford Radioactivity in Salmon Spawning Grounds', may give the real reason why. It says 60% of the Hanford Reach riverbed is contaminated by Hanford's radioactive waste coming not from plutonium production as expected, but as a by-product from a still secret Thorium to Uranium-233 process whose toxic pollutants are harder to detect. The list of trace elements revealed in sediment samples is almost endless. Names like ceasium 137, europium 152, cobalt 60, potasium 40. The deadly toxic cocktail mix doesn't end there.

While Washington State tries increasingly frantically to reassure its citizens that water quality in the Columbia is the best there is and there's no danger from eating the river's fish, the Indians who have come to know the land and river well over generations, are far less sure.

"The ground is moving," they say. "Not much, but we can hear it. The earth is stretching. It is crying. It is tired of what the White Man has done and is trying to rid itself of his poisons We think a huge earthquake is coming." But what do they know? And who would listen?

To environmentalists some of it may sound redolent of British scientist and environmentalist James Lovelock's "Gaia Theory". It takes its name from the Greek goddess who was mother of all. The theory itself sees the Earth as a living, organic body continuously adjusting itself to cope with what Man does. Over time the planet will recover from Man's actions. Time is the one thing the planet has plenty of. So far 4.5 billion years has passed compared to Man's puny couple of million years of existence. And a 100,000 years to Gaia believers is nothing. On time scales this long the planet will undoubtedly recover. But in the process Man may no longer be around to see the result. At best he may long have been buried under the ice.

What it said was nothing that couldn't be contained, was Coleman's first reaction, rather than an alarmist "Oh, Shit!" which he suspected others might think. The more disturbing fact was that the article had appeared at all.

His car pulled smoothly to a halt outside the Eisenhower Executive Office Building (EEOB). But the article had succeeded in putting him into a sombre mood; so sombre he didn't wait for the door to be opened by his government chauffer, but instead grabbed his copy of *The Post*, jumped out and sprinted up the steps.

The EEOB was a meticulously restored French Second Empire-style building, once close to demolition in 1958 before being saved as a National Historic Landmark. It lay on the corner of Pennsylvania Avenue and 17th Street NW, adjacent to the West Wing of the White House.

Originally built in 1871 the EEOB housed many of the president's executive staff, including the office of the vice president. The West Wing, next to the White House, housed those involved in the day-to-day activities of the Administration, such as the presidential Chief of Staff, speechwriters and Press Secretary. It was where Coleman's first meeting of the day would be with the president, and was scheduled to last no longer than thirty minutes – less if possible.

The office of Director of National Intelligence for which Coleman had recently become responsible was recognised by Washington cognoscenti as a bed of nails. Its remit was to coordinate a US intelligence machine, estimated by *The Washington Post* to employ 854,000 people with top secret clearances, across 16 main federal agencies, 1,271 other government agencies, and 1,931 private companies – all located on 10,000 different sites across the country. Internal political machinations, historical agency fiefdoms and inter-agency rivalries were the three principal reasons why it was never going to work seamlessly.

Dexter Coleman's plan on arrival at the EEOB was to use its unobtrusive linking tunnel with the White House, so that as few people as possible would see him enter that morning.

CHAPTER 7

The Roosevelt Room was opposite the Oval Office and it's two marine guards, on the left-hand side of a wide and oil-painting hung corridor. Chief of Staff Clifford Wurzburg was already seated at the end of the long repro antique table with his back to the fireplace and the picture over it of a horse-mounted Theodore Roosevelt as he liked to have imagined himself.

'You've seen *The Post* this morning?' was Wurzburg's first comment as he entered.

'Sure,' said Coleman, flinging his copy on the table.

'It shouldn't spook us. They know nothing. It's just routine stuff like they've been writing for years. But the president wants to talk about it. *The Post's* White House correspondent warned him it was coming late last night. I think the president sees it as a wake-up call to the troops. That means us.'

Coleman wasn't sure what he was talking about.

'Where's Hoop?' he enquired as he took off his heavy, charcoal-grey overcoat and laid it neatly over the back of a red leather-covered chair.

'He's coming. Or he'd better be, and soon,' replied Wurzburg as he checked his watch.

Wurzburg was a large and powerfully built man in his mid-fifties, a former Caltech engineering graduate with a fierce reputation for running a slick and efficient Administration. He'd already served with one president ten years before as a Special Advisor, before becoming Secretary of Transportation with the previous Administration. It spoke volumes that he had been retained by the current president and placed in overall charge.

The door to the Roosevelt Room opened again, revealing the portly figure of Hoop Toberman.

'God, it's cold out there,' he said, wheezing for breath. Hoop was an avuncular-looking sixty-three-year-old from Elizabethtown, Kentucky, where for five years he'd been state governor. Now, as the current Secretary of Energy, he controlled a budget of nearly $40 billion with over 100,000 federal and contractor employees, making his department the eighth largest recipient from the annual $4 trillion federal budget. It was still a long way short of the Department of Defence budget that during the current year fell just short of $700 billion.

There was no time for any small talk as President Juan Sanchez entered close behind and took the seat at the head of the table, closest to the door. In unison the three men said: 'Good Morning, Mr President', and stood to attention.

'Be seated gentlemen. Let's get this nut cracked.'

Sanchez was small, fifty-one, with uncharacteristic, piercing blue eyes for someone of otherwise obvious Latin origins. It was early days for his presidential term of office, so he knew well that if you wanted to make changes then you did them early. If you didn't then the system would bog you down. So he was a man in a hurry.

Sanchez quickly followed up his introductory statement with: 'I got a heads-up on *The Post's* story late last night, which is why I've called you all here. I've taken it as a wake-up call. So from here on in there's gonna be more meetings like this one as the DEEP EARTH deadline approaches. We may have to draft other people in to get the job done, so let's be aware of it. Notwithstanding that we have to maintain the greatest need for secrecy, which is why Dexter is here. Let me say informally, this is the first meeting of the newly formed DEEP EARTH committee. As from today, October 1st, it takes priority over anything else you might be doing, at least for the next 151 days.'

'This is how we're gonna do it. We give DEEP EARTH our best shot and we have to have it either work or fail by end of February next. Beyond that, if it hasn't worked, we have a conventional back-up plan that calls for us to evacuate twelve million people in the Pacific northwest. Our best guesstimate is an earthquake is going to hit us in June of next year. So we need our armed forces, National Guard, Homeland Security and the rest to be on standby for that date without spreading alarm and despondency to people who live in the area.

'For DEEP EARTH we're allowing three months plus one month contingency. The contingency is the time between beginning and end of February. We have to have all our calculations in place and be ready by the beginning of February so we can act on it within 30 days.

'The bottom line is if DEEP EARTH fails, and heaven help us if it does, we start our evacuation plans beginning of March! Yep, that's how long we've got, I've been told. Get used to it.'

He paused for a few seconds to let the information sink in. 'But for DEEP EARTH alone we've got just 151 days. Let me just say that again … it takes priority over anything else you've got – or get – on your desk for those 151 days. That's just five months. That takes us to the end of February next.' He stopped talking as he allowed the deadline to sink in. No one said anything. 'It's not long. The quake itself is scheduled to hit us by the end of July. Take one month off that for safety. Then knock three months off that, which is the time the Department of Homeland Security, with the backing of the Department of Defence, tell me it will take to evacuate twelve million people from along the Columbia.

'Since Hurricane Katrina, which wrecked New Orleans, Homeland runs continuous computer simulations about how long it will take to evacuate any given area. We gave 'em the scenario we actually could have in Washington state. Three months to evacuate is what they said. So that brings us to a deadline for our fuzzy end of the stick to, let me say yet again, the end of February. It's no time at all considering what has to be done.

'Hoop here is gonna fill you in as to where that time is gonna go.' The president stopped for a moment and gave a long glance in Toberman's direction before adding, 'That right, Hoop?'

'Yeah, that's right, Mr President,' Toberman replied, before the president continued: 'Following *The Post* article, I got Hoop here to send me over the latest report on seismic activity at the northern end of the San Andreas fault line. I take it you've all had that?' He paused and looked around. Nobody said anything. 'Good. I'll take it as read.

'As you know, the San Andreas fault runs up from south of Los Angeles for 810 miles and ends at Cape Mendocino. That's the good news. The bad is that it connects with a sister fault that affects the Cascade mountain range: Mount Rainier, Mount Hood, Mount Adam and Mount St Helens. They all sit on what's called a subduction zone. The US Geological Survey tell me this is where the Pacific tectonic plate moving east meets up with the static US continental plate, before it dives down underneath it and heads towards the earth's centre. In the process, it causes fractures in the earth's crust, along which we get lively volcanoes.'

The president paused momentarily to make sure his words were registering with his small audience. He continued. 'So far we've been lucky. Only Mount St Helens has blown its top in recent times, in 1980, if you remember.' He paused for effect. 'Well, gentlemen, our luck's about to run out.

'The fault line under the Cascades wanders off under Washington state, below the Columbia River, and then under the 600 square miles of Hanford Reservation. You all know what happened there.' It was a rhetorical question. 'If you don't, then read *The Post* article: it's why we're all here. It's why we've got 1,000 contractors busy right now, digging the necessary DEEP EARTH tunnels. The plan is to use a small underground nuclear explosion to suck out all the shit we have stored there deep underground. Come any quake, its effects on releasing any nuclear waste through opening up of cracks into the river and ground water will be minimal. We hope. What we're waiting for is precise data on where and how deep some of those tunnels need to be. We get it wrong and we may well make things a whole lot worse. Hoop is gonna update us on that.

'So, gentlemen, what we're trying to do is head off at the pass the results of a projected earthquake. It's not the earthquake itself that scares the hell out of me, although that might be bad enough, it's what it will do to Hanford and its radio active waste, and what that will do to the folks who live up there.

'Remember what happened to the Japanese nuclear reactors at Fukushima in 2011? That's kid's stuff compared to what might be heading our way. It's all hands to the pumps if I'm not to semi permanently evacuate twelve million of our citizens who live up that way, along the Columbia River, including cities like Seattle, Pasco, Kennewick, Richland, North Bonneville, and a hundred others I can't remember. It's all gonna be as radioactive as hell if we get this wrong. So failure is not an option.'

With deliberate wry humour the president added, 'Worse, I'm the guy who'd have to order the evacuation. Even though it's not my fault, I'd be the guy they'd all hate. So, Hoop, give me some good news on the latest predictions with our earthquake. We're all aware, too, that time is not on our side.

'Now here's the joker in the pack. Helping us is some Iranian geophysicist we've recruited recently but who knows all there is to know about predicting earthquakes and volcanoes. He's been a success in Iran with his predicting capability. But the mullahs kicked him out. That's when he fell into our laps care of the British. Start first, Hoop, with if I need to worry about this morning's *Washington Post* article. Whether there's anything new I need to know about Hanford.'

'Nope, I don't think so, Mr President,' said Toberman. 'Mount St Helens, ninety miles north-east of Portland, Oregon, is our watchman. It's at the northernmost extremity of the San Andreas before it veers westward out into the Pacific, where it's no longer our problem. It last blew big May 18th, 1980, at 8.32 Pacific Day Time. It devastated 200 square miles around it. Just in case you've forgotten, it was also the most economically destructive event in our history. It'll start blowing its top again ahead of the big one, the earthquake, so our Iranian friend tells me.'

'Even then, we have a further back-up with Mount Rainier, fifty-four miles south-east of Seattle. This mountain is really big. At 14,411feet, it dominates the skyline, and is the biggest in this country. Our Iranian friend says it's a massive stratovolcano and, when it goes, it's really gonna go. It's considered one of the most dangerous volcanoes in the world, and is part of the fault line that runs on up Washington state, and under the Hanford Reservation and the Columbia River ...'

'I saw *The Post* this morning,' Coleman said.

'I guess we all did,' responded Wurzburg tartly.

Toberman took up the story again. '*The Post* article is a drop in the ocean if Mount Rainier starts to give us trouble. The good news is that it might, and I say might, give us a warning before Hanford is affected. What *The Post* was on to is the permanent leaky state of the holding vessels we've already got up there. They've been leaking for years, ever since they were built in 1944/5, along with the original plant back in the early 1940s. It's gonna get a whole lot worse as we get nearer the predicted event time. Then fissures will start opening up all over the goddamn place, which is when the toxic and radioactive crap, stored there for decades on the Hanford Reservation, will begin emptying into the Columbia. Then the shit will really hit the fan, believe me.' He smiled as he realised the unintended joke he'd just made. 'And, gentleman, I wish it was "shit" because that's nothing to what's really down there. One cupful of it could kill a hundred people in seconds.'

'God, it could be a disaster through generations,' Coleman muttered. 'All the radioactive waste resulting from seventy years of plutonium production.' Dexter pulled a face. 'And it covers an area of 600 square miles.'

Wurzburg added, 'The result of producing enough plutonium necessary for building the 60,000 nuclear weapons we held on to for our pre-treaty, pre-peace dividend stockpile!'

'Let's not start panicking, gentleman,' Sanchez advised, sensing the trepidation his team were beginning to feel about what lay ahead. 'It's why we have DEEP EARTH.'

The president turned to Toberman again. 'Tell us about the Iranian and how he's helping us.'

'As you know, Mr President, he's a world authority in vulcanology and earthquake predictions. Over the last ten years he's got better and better at it. He's usually right with what he says ...'

'Usually?' the president asked.

Toberman ploughed on. 'It's as good as anyone can be, Mr President. He predicted the Bam earthquake in Iran, then the one in Turkey that threatened Istanbul. He also successfully predicted the 5.2 magnitude event in Auckland, New Zealand. His predictions allowed time for people to get out. He saved thousands of lives. He's an expert because Iran is one of the most tectonically active countries in the world. It sits on three fault lines, and Iranians live with the results every day. Earthquakes are a way of life there.' He then added grimly,

'And death. But that's where the experts are, Tehran University. It's where he taught. He was the professor. He's taught everyone who knows anything, about anything in this area.'

'We're lucky he fled his native Iran and went to the United Kingdom first. Our friends in the British Secret Service tell us he was at some London university for four years before coming here. He has family in Baltimore: his wife's parents. The father-in-law got out years ago, before the trouble with the mullahs got outta hand. Took all his money out too. He's a multi-millionaire, maybe even a billionaire. But money didn't bring him health. He's a sick man; something to do with his breathing. Emphysema, maybe.'

'Emphysema?' echoed Wurzburg.

'He's not got long to live,' added Hoop. The Energy Secretary then continued: 'Our friend's computer simulations have been very good at predicting earthquakes, and he's done a lot of new stuff on the San Andreas fault line and the Cascade Range since he began teaching at UC Berkeley, in San Francisco, three years ago. The slightly good news is that although Mount St Helens, Mount Rainier and Hanford are all connected, it doesn't mean they all have to blow their tops at the same time. It could just be St Helens, or Rainier, or Hanford ... or maybe two of them. It could be that activity in one will relieve the pressure on the others. We just don't know. We gotta do a lot of praying, gentlemen, that there is a God, and that He helps us. What we do know is that we could live with just the aftermath of pure volcanic or earthquake activity. It would be tough, but we could do it. Hanford is different. Its effects will last hundreds of thousands of years. We might need to permanently evacuate a part of the United States.'

'But what's this Iranian up to now, Hoop?' the president asked impatiently. 'Where's he got to? We need his input for those damn tunnels. Without those we'll be drilling blind.'

Toberman tried to be reassuring, although he didn't feel it. 'We've just given him access to the computer facilities at Los Alamos, the most powerful computing facility we've got.'

'What's this guy's name?' asked the president.

'Pahlavi,' Cliff Wurzburg answered. 'Robert Pahlavi.'

Hoop Toberman rejoined the conversation. 'He kept saying he needed more power and yet more computing power. He knows that after Los Alamos we're outta road.'

'Is he gonna meet the deadline?' Sanchez asked.

'He'll meet it,' Wurzburg cut in again. 'He'll have the data on the tunnels when we need them. Hoop and I have worked out a schedule for when that will be. February 1st is the actual deadline our Iranian friend gives us, plus or minus ten days, he says. So that means February for us, at the very latest. But we need to have our evacuation back-up plan ready for action a month before that, in case DEEP EARTH proves the world's biggest fizzer and buries us all. So that gives us end of January 31st, beginning February again.'

Sanchez continued: 'So time is tight.'

'151 working days, Mr President,' said Wurzburg, reiterating what they'd already heard from the president a few moments earlier. 'It's our latest, best estimate. Five months from here on in.' He looked grim.

'This is a delicate balancing act we have here,' continued Sanchez. 'We don't want anyone panicking. At the moment, I think we're slightly ahead of the game.' He looked at Hoop again with a piercing gaze before continuing. 'If our Iranian friend and his team, plus the others from the US Geographic Survey, get it right, then DEEP EARTH will swallow it all.'

'The cavern created by our two-megaton nuclear charge will collapse soon after its creation, sucking in all the crap we've been storing for years. The natural movement of the subduction zone will then carry it towards the earth's centre, where it'll be perfectly safe. Pretty neat, huh? We turn the subduction zone into our friend rather than our foe.'

'That's the theory,' said Wurzburg. 'If it works.'

'Yeah. That's the theory,' said the Energy Secretary.

'It'd better work,' said Sanchez, 'otherwise there's gonna be more than egg over my face. I'll be looking for a new job in a parking lot.'

'There's still gonna be an earthquake,' he continued, 'but Hanford will be safe. But for another six months we've got to keep the lid on it. Maybe forever, until all the danger has passed.' He turned to gaze at Coleman. 'I take it we're keeping a careful eye on our Iranian friend? We don't want him jumping ship just when we need him most.'

'I have my best men on it. They babysit him morning, noon and night,' said Coleman. 'According to our British friends, he's clean. They watched him to see if the Iranians would try and blackmail him when he was there. He still has a son in Tehran who's a surgeon. They might've put pressure on him to pressurise his father. There was no evidence of that, the British said.'

'Okay, gentlemen ... until next time.' With that everyone knew they'd been dismissed and the president was on to his next problem. It was only the tough, unsolvable problems that ever came to his desk. The rest were someone else's responsibility. Cliff Wurzburg saw to that.

The meeting had lasted twenty minutes. Already the president's Principal Private Secretary, one of six, was waiting at the door for him. Without a word President Sanchez rose and followed her out.

Cliff Wurzburg motioned the other two to wait. They sat down again.

'Just in case you're wondering where we are, constitutionally, with all this ...' – he paused a moment to let the words sink in – 'the Cabinet meets once a week most times, under the Constitution's Article II, Section 2, which states merely that the president "may require the opinion, in writing, of the principal officer in each of the executive departments, upon any subject relating to the duties of their respective offices". The Constitution does not say which, or how many executive departments should be created. In other words, we need only to inform them when the president thinks it's necessary. We have a good precedent for when that should be, particularly when it comes to matters of national security, as DEEP EARTH clearly is. When Vice-President Harry Truman was sworn in, after Roosevelt's death in 1945, he had no idea the Manhattan Project existed. And he was the guy who a few months later had to authorise the A-bomb's use on Japan!'

'Yes, but ...' stuttered Coleman, before he was cut short.

'There are no buts,' interjected Wurzburg. 'We have twenty-two people in the Cabinet. How many of them do you think would keep the lid on DEEP EARTH? How long after that before there would be panic? The question is: What do we know, and when will we know it? Until it's been proved, we know nothing. We all know San Francisco sits astride the San Andreas and is gonna fall into the deep blue sea sometime soon. That could be tomorrow or in a thousand years. With DEEP EARTH we're way out on a limb, to the tune of $4 billion already. It would look kinda foolish if it was unnecessary, and even more desperate if we went public, and got ourselves involved in endless public debates, cut short because we were proved right. We'd be damned if we did and damned if we didn't. But someone has to make the decision now. And that buck stops on the president's desk. You understand?'

Wurzburg paused. There was nothing, Coleman thought, he could add. Hoop Toberman thought likewise. It was a tough call, they both murmured in unison.

'Okay,' said Dexter. 'Glad we got that out of the way.'

'Sure,' replied Wurzburg. 'Now you both know.' He rose and headed for the door. 'And we gotta have a back-up plan just in case DEEP EARTH fails. Have you thought how big an operation that might be? Trying to evacuate twelve million Americans, on the quiet?' It was a rhetorical question. He waited for them both and for his walk down the corridor back to his own office.

Dexter and Toberman looked at each other with blank faces and said nothing as they collected their belongings and prepared to follow.

CHAPTER 8

It was a Saturday morning in early November. There was little movement along the secluded cul-de-sac that linked cottages and farm buildings on the edge of Ripley Common. Writer H G Wells had given the common some notoriety when in his book "War of the Worlds" it was where the Martian tripods are first seen advancing towards London. Now an occasional and often elderly track-suited runner could be seen jogging slowly in belated attempts at keeping fit.

The first of the cottages was three-bedroomed and built in the late nineteenth century. It was white-clapboarded, with a recent clapboarded extension skilfully added to match the rest of the house.

'So let me see if I've got this right,' said Katrina from her seated position on a tattered leather armchair, laptop on her knee, books, newspapers and assorted research documents strewn on the floor around her. 'In terms of you taking on the government, with its army of lawyers, assorted heavies and billions of pounds to spend, you think the best place to start is with a couple of bent coat hangers and an ancient book nobody believes in, written by someone long-dead called Harold Watkins?'

She was picking up again on a short conversation she and Jonathan had started in bed the previous evening. Now it was the morning, eight hours later. 'Don't you think this is going to be a little short in the ammunition stakes, considering what you plan to do? Not that I understand what that is.'

'Sure,' said Jonathan completely unfazed, his mouth half-full of toast as he spoke. 'I wasn't thinking of metal rods so much as untwisted metal coat hangers,' he continued matter-of-factly. 'And one's ability to out-think the enemy,' he added with slightly more emphasis. Then as an afterthought he added, for no particular reason, 'Which is not something Chinese military philosopher Sun Tzu never said in his influential treatise, *The Art of War*, published two thousand years ago. I think people sometimes think he did say it, on the basis he said everything else worth saying on warfare.'

Katrina shook her head to clear long strands of black hair out of her eyes. She could see Jonathan was in a surprisingly good mood for early morning, which was unusual. She took heart from it.

'And this is the book that's got all the information I need.' He picked up a small and old-looking book with an orange-brown cover. 'It's called *The Old Straight Track: Its Mounds, Beacons, Moats, Sites and Mark Stones*,[1] he added amiably. 'First printed in 1925, by a Mr Watkins, as you correctly observed, then reprinted repeatedly over the last nearly one hundred years, with its last edition appearing in 2006. Which means people are still interested, irrespective of how often scientists debunk its subject matter ... but never quite completely, funnily enough.'

Katrina envied the way Jonathan was able to recall facts and figures with consummate ease. For her post as research fellow she had to memorise things the hard way, through constant repetition and familiarity. Jonathan just had 'one of those brains'. She liked to think she evened the score by being a consistent early riser so, several hours before Jonathan had appeared, she had been up, poring through several books at once, taking notes, and using Google to find more information on late Roman Britain AD 410 to 550: the period covering the end of Roman rule in Britain, up until what had once, mistakenly, been called the Dark Ages. It was a period when the Saxons had been invading the country to be resisted by a shadowy and legendary fighter called King Arthur.

---

[1] *Abacus (New Edition) (1st January 1988)*

But did Arthur ever exist in reality, as some insisted?

'The vicar's asked me to take the family service tomorrow morning,' Jonathan shouted over to her. 'He's got family visiting from Wales, so his hands are full. So I thought I'd give the Parishioners a good taste of fire and brimstone. A sermon on "forgiveness" I think would be appropriate, considering the bastards I've been dealing with this week.'

Katrina said nothing, but continued with her preliminary research work.

'Maybe I could look at Our Lord's last words on the cross: "Lord, forgive them, for they know not what they do." Think of it. Being nailed to a cross is not a pleasant way to die. It takes time; a long time. The weight of your body is hanging on the nails. You're thirsty, hurting, suffocating because of the way your body hangs. It's a terrible, terrible way to die. It's torture. And as he hangs there in horrible pain he asks for forgiveness for his tormentors. Yeah, I like the sound of it. That's my sermon for tomorrow.'

Katrina still said nothing. She knew this is what he did at the beginning of each day: use her as a sounding board for the things on his mind; but she'd learned, too, to be a silent witness.

It had been 10.00 a.m. when a still bleary-eyed Jonathan had first appeared that morning. He always slept lightly, meaning they often went to bed and rose at different times. It was the way their respective metabolisms worked. When Jonathan did rise he didn't do much, not straightaway. He had to come to terms with a new day slowly, so his first actions were always of displacement therapy, Katrina theorised.

For half an hour he'd pottered about, boiling two eggs, while making two slices of buttered toast, one of which he neatly cut up into soldiers for dipping into his eggs. The other slice he coated thickly with butter and Baxter's thick-cut marmalade his mother still sent him from Scotland. Along with a large glass of freshly squeezed orange juice with ice came freshly brewed Italian coffee, made in his own and prized cafetiere which Katrina had once bought him for his birthday. Evidence of his 'work' remained everywhere; he was not a neat person. Katrina thanked the gods that, so far, this was something he only undertook at weekends or on public holidays, otherwise it was a cup of coffee before he disappeared out of the door for his drive to Berkshire. He was not known for his punctuality!

As always at weekends, and first thing, his hair remained unkempt, his face unshaven, his tracksuit stained. Overall, Katrina thought as she had watched him, he had the overall appearance of an unmade bed.

She was used to his morning ritual and knew better than to disturb him until he was good and ready. So after her opening remarks she tried not to let the general commotion he had caused distract from her reading up on material she might use for her projected new television series.

Finally, after Jonathan's second egg and slice of toast had been washed down with coffee, she appreciated he was ready to talk about his resignation, and what he now planned to do in the wake of it. Katrina was a woman of guile. She knew how to get the best out of her husband. To further the conversation she decided on a different tack. Without saying anything she got up from her armchair, walked over to him, and gave him a 'good morning' kiss on the forehead as she placed her arms around his neck. He looked up at her as he sipped his coffee.

'It's been a terrible few days for both of us,' he volunteered.

'It's not been great,' she agreed. 'How was Ben yesterday?' She asked as it had been Jonathan who'd looked after their son for most of the day. Friday was her day at university, 'putting in a face', as she would say. He understood. Normally, Ben would have been at his special school, or in the care of either Dolores or Adrianne, both fully trained paediatric nurses, depending on whose shift it was. But yesterday, Ben had missed school because of a mild 'episode', and Jonathan had, unusually, seen a lot of him. In addition to the nurses there were regular visits from the special needs section of the local Health Authority which, Katrina and Jonathan admitted, had so far been exemplary.

'He was fine; like nothing had ever happened to him,' Jonathan replied.

'That's the way it is, and that's the way it's always going to be. It's one of the hallmarks of his illness,' Katrina responded. 'It won't get any better, is the bad news. The good is that it won't get any worse.'

'Yeah, I know. It still takes *me* some getting used to,' he said. 'Thank God we've got Dolores and Adrianne, though I hate it because I think we should be able to do it all between ourselves.'

'Me too,' Katrina replied sympathetically. 'Maybe it's normal to feel guilt. But with our jobs we just wouldn't have coped,' she reminded him, ' and however much we wished we could. It was to get the money for the support we needed that you did the deal you did, with your old employers. Look at all the aid we've got out of it, including being able to afford the nursing help. The extra cash I've earned from television has helped, too. Think of the elevator we had installed to take him and his wheelchair up to his bedroom; a car large enough for getting his wheelchair in and out; the special bathroom for him. I don't need to go on. We're lucky to be where we are. Think of the poor sods who aren't so lucky. Many of them have to give up their jobs to cope and live on the breadline.'

Katrina untangled her arms from round his neck and went back to sit back down at her chair.

'I know, I know,' Jonathan continued irritably as she sat down. 'We're the ones who've got it easy too, while it's Ben that's got it hard, and for the rest of his life.' He then realised he hadn't seen his son that morning, and asked where he was.

Katrina replied. 'Dolores has taken him shopping. You know how he likes to get out. She's gone to Sainsbury's, the big one down the A3.'

'Will she be okay if anything happens?'

'Sure, she'll be okay. She and Adrianne are both very capable. But who knows until it happens? Though I think the risk is low, with his last episode being only a day or so ago. However, he could have another episode tomorrow.

'Funny how things work out,' she ploughed on. 'And something I didn't have time to tell you was that Caedmon, the TV production company, have this morning confirmed they want me to do another TV series. Don't you think that's good? Go on! Be pleased for me!'

Jonathan smiled across at her. 'Jeez, that is good, darling. In fact, it's more than good. It's very good.' He paused a moment as he seemed to lose interest. 'Sorry, I'm still a bit preoccupied. I've got this whole fusion thing going on my mind; whether I did the right thing.'

'Don't ever doubt it,' she said reassuringly. 'So, I've listened to your thoughts on tomorrow's sermon, now what about us? What about your Alfred Watkins and his ley lines and coat hangers? How do these things fit together to form a plan? How are we going to beat the buggers? You quoted Sun Tzu and his *Art of War* to me earlier. Well, what about Karl Marx and his *Das Kapital*, where he said, "the committed few will always overpower the uncommitted many"?'

Jonathan sprang to his feet as if her comment had struck a chord. He walked the few steps to where she was sitting. It was his turn to place his arms around her neck. As he did he said, 'As my very closest advisor, you know that old adage "show, don't tell"? If your TV and university can afford to let you go, I want us to go to Cornwall for a few days, take Ben too, and Dolores or whoever is available.'

Katrina gently held Jonathan's left hand. 'Sounds fun. What about the good old British weather?'

'Never the wrong weather, just the wrong clothing,' he responded positively. 'I've been there often enough at this time of year. Even in late December it can be pretty warm. Trust me.'

'Is this just a plain vanilla holiday? Or is there more?' she asked suddenly suspicious.

Jonathan laughed a short and humourless laugh. 'I can't fool you can I? Of course there's more. This is where the book and coat hangers come to the fore.'

'I'd like to go in, say, a couple of weeks? From memory that's when BRIGHTSTAR will begin the first of its power run-ups. Although it'll be at one of the lower of the power settings, it should still be enough to make my coat hangers twitch.'

'How so?' she asked.

'The ley lines given so much credence by our friend Watkins,' he answered. 'Although ridiculed by academia, not so according to my theory. There may well be such things …'

'Okay,' said a puzzled Katrina.

'… but there may be more. If my extended theory is correct then along the beam and directly underneath it there will be a residual effect similar to what they claim to be from ley lines or water dowsing. If there is such an effect, close to where we'll be staying, then hey presto!'

'It's not going to shrivel us up, I hope,' Katrina responded, 'otherwise you can go on your own.'

'No, no. The effects, if there are any, will be tiny.'

Katrina swivelled her head to look up into his face. She hadn't understood much of anything he'd said. She was used to that. But the bit about going to Cornwall for a few days was good news.

CHAPTER 9

'How far have we got?' asked President Sanchez in a firm and authoritative voice. 'The clock's ticking, let's not forget. Almost a month has gone past since our last meeting. It's nearly December. We have just over sixty days left to our February 1st deadline, come what may. Let me hear about progress, what we've done, and, hope to Christ, no hang-ups.'

This meeting, unlike the last, which had been held in the Roosevelt Room, was being held in the Oval Office. Since Nixon's time, it had been wired to tape all conversations. Dexter Coleman, Head of the National Intelligence Directorate, Energy Secretary Hoop Toberman, and Chief of Staff Cliff J Wurzburg, were sitting in three chairs facing the president's desk. From there, they were able to look past him and out through the thick, green-tinted, bulletproof glass, across the Rose Garden, to the needle-like obelisk of the 500-feet tall, white Washington Monument that lay beyond.

'Mr President, everything is on track. So far no hold-ups,' said Toberman, who then proceeded to reel off a long list of figures relating to engineering progress at Hanford site, movements of radioactive materials from around the country to Hanford, and of the evacuation plan in case it all failed. It was left to Dexter Coleman to reveal a possible problem.

'We thought we might have had a problem, Mr President,' said Coleman. 'One of our field agents, involved in watching and protecting our Iranian friend, Pahlavi, was murdered a few days ago.'

'What? Killed? Did I hear that right? Jeez. Tell me more before I send for the two marine guards posted outside my door,' said a clearly agitated Sanchez.

'We don't think it was anything, Mr President. It was coincidental,' added Coleman hastily, trying to dampen down the situation. 'On a par with a traffic accident, we think.'

Wurzburg quickly jumped into the conversation: 'The reason I didn't bring it to your attention before now, Mr President, is that we investigated it, and we think it was a case of one of our agents being in the wrong place at the wrong time.'

'Go on,' said Sanchez.

'Our man was watching Pahlavi at some bar in Sausalito,' continued Coleman, 'when evidently he was whacked by some smackhead looking for money for his next hit. He chose our man. We have nothing else to go on to indicate it was anything else, nothing more suspicious. No link with anything. It was even amateurish in its execution.'

Then the Chief of Staff cut in: 'Mr President the only thing suspicious we have is that our agent was killed by someone toting a Glock 17 judging from the 9mm bullet we recovered. It happens to be the favourite firearm used by the world's security forces. But in the US where the FBI estimates there's at least one firearm for every man, woman and child in the country, there's bound to be a lot of Glock's in the mix.'

'What about Pahlavi?' asked Sanchez.

'Nothing,' said Coleman. 'It's business as usual for him. He's coming up with the goods and our engineering guys at Hanford are implementing them. From his point of view maybe nothing has changed. But we have nothing to go on to say he isn't right. Needless to say, we've geared our men up and warned them to be extra vigilant.'

∞

The fluorescent lighting running past Ben's private room at the Atkinson Morley had been dimmed for the night. Katrina was still with him, just as she had been since the late afternoon; reassuring him, soothing him, telling him there was nothing to fear. Although the hospital was a recognised world leader in neurology, even as she was saying it she knew that Ben could not understand. But she continued to hope against hope that one day he might. She prayed for as much, and that Ben would not have to spend much more of his young life lying in rooms just like this one. This time she felt hopeful for some reason; that this time it might be different.

The hospital had a new 'Nuclear Imaging' department, she'd been told, and with it was trying a new diagnostic technique that one day might lead to Ben becoming 'normal'. After so many previous false dawns she hardly dared consider it. All she could think about was that Ben had already endured so much in his short life.

Occasionally, a nurse or a doctor hurried past on an errand to somewhere else. A small table lamp beside Ben's bed was turned off. A red indicator light on a low-light-level CCTV camera mounted high on a wall winked intermittently. A researcher in another building, a quarter of a mile away, watched the amplified, grainy-green images it transmitted.

Every so often, a twenty-five-year-old University College postgraduate made an entry into a notebook. She had been there since 7.00 p.m., with her shift finishing at 2.00 a.m. Her co-researcher, a research neurologist, would be taking over the graveyard shift at 2.00 a.m. and finishing at 8.00 a.m.

Before leaving for the night, consultant neurologist Dr Hugh Orum had spent time explaining to Katrina that the cumbersome helmet Ben was wearing was a health-care spin-off from nuclear weapons research at Los Alamos, America's most prestigious research labs. 'They pioneered development of the world's first atomic bomb,' he had said. 'And subsequent to the international ban on all nuclear weapons testing, they now do it all on the world's most powerful super computer, which models what the effects of different nuclear weapon designs would be. Their super computer is so powerful they have spare capacity that they let qualified hospitals like ourselves use.'

'Apart from five US universities, the Atkinson Morley is the only organisation outside America to be given such access, along with a link to their high-speed data network, linking us to the lab's computing centre at their Department of Nuclear Medicine in New Mexico.'

Orum had gone on to say that Ben's headwear was called a SQUID, short for Super-Conducting Quantum Interference Device. It alone was worth two million dollars, and was an American invention. Embedded in it were millions of sensors sensing various levels of Ben's brain activity. These were then transmitted via a radio link, to a remote researcher capturing it all before onward transmission to the Los Alamos computing centre, somewhere in New Mexico Orum believed.

This centre had the power to compare those cleaned-up brain signals with a library of other, similar stored and pre-recorded signals, derived from hundreds of 'test' patients. The centre had viewed millions of photographic images in order to see what brain signals resulted from what image. Millions of specific brain patterns were then correlated with millions of specific images. New images with new signals were being added all the time. In practice, it meant Ben's brain signals might eventually be turned into near-real-time moving pictures. In short, medical researchers could see what Ben had been thinking. At least, that was the theory. Although results from it had so far been promising, it was nevertheless still at an early stage of development.

The Atkinson Morley was privileged, Orum had said. Then, before he had left for the night, he had said with a wink that Ben should also feel privileged at having the use of such expensive equipment. He had smiled at Katrina in a kindly way, squeezed her hand, and reassured her that Ben was in good hands. She had no need to worry. Then he had turned and gone. Katrina watched him go. She realised she liked him. He was a 'good' man, she thought.

∞

Ganhumara looked at her warlord for a moment and smiled. She had long, dark hair surmounting eyes containing the pools that were her violet-coloured irises. He smiled back and held out his hand. Together, they left their home on the highest point of their island fortress that was not really an island, for it still adjoined the mainland by a narrow neck of land. They walked out into the coolness of the approaching evening. The sun would soon slip below the distant horizon, cooling the day that had been long and hot. It had made it especially hard for the long lines of men unloading the heavily-laden corbita, the Mediterranean cargo vessel that had arrived a few short hours earlier, loaded with eagerly awaited produce from the old world of Rome and its empire. This would be exchanged for exquisite Celtic ornaments, along with ingots of tin, lead, copper, silver and, occasionally, some gold. Above the incessant roar of the sea a hundred feet below, they could also hear the voices of men going about their back-breaking work. Tonight, they would rest and feast, and drink and swap stories with the corbita's sailors about their respective homelands. Then, tomorrow, the ship would sail westward, further along the coast, before traversing the inland sea to exchange the last of its Mediterranean goods at Ynys Witrin, the island home of Myrradin the Wise.

Ganhumara and her warlord took the most westerly route along the narrow path that bisected the island, before it led eventually to the headland. As was their custom, they looked out over the calmness of the ocean, to the sun setting in the far west, the birds wheeling and soaring on columns of warm air sweeping up the cliff face, and interpreted cloud patterns in terms of what tomorrow would bring. And they looked across the shallow, curving bay to where Morwenna lived, and recalled her frightening stories of capture by Irish raiders, of life with her father, King Brycheiniog, who ruled that part of Wales now called Brecknockshire. But her years of capture had deranged her mind so that she saw frequent visions of Christ, strange drug-enhanced images not of this world to which she would chant and dance. Morwenna would claim control of the ancient lines of force that, she said, flowed through her home, before connecting with all centres of belief, including their own island home at far-off Ynys Witrin to the east.

As they walked, they passed close to where their island joined the mainland, to where men still toiled up and down the path leading to the beach and the anchored corbita.

'The feasting, drinking and storytelling will be raucous tonight,' the woman said, as they walked hand in hand.

'Aye, that it will be,' replied the man in Latin. 'We will discuss business too, and the way of the empire. And the merchantman from the corbita will ask me about Britain, and whether we would welcome reoccupation.'

'You know that?' Ganhumara replied with surprise.

'It is common knowledge. Perhaps the only one who does not know it is the ship's merchantman.' She caught the tinge of irony in his voice.

'Would you welcome reoccupation of Britain?' she asked. The man seemed ill at ease with her question, and she sensed it. 'Is there anything wrong?' she added quickly.

'A messenger arrived earlier today. We are to expect attack from a Saxon raiding party, maybe numbering many hundreds. They will come from the southeast, from among those Saxons already settled as peaceful farmers, who will be told either to fight or face being put to the sword. They will have little choice.' He grimaced as he said it. 'You know the Saxons are savages, destroying all they do not understand.'

'You have fought them many times,' she said. 'And you have won. The people are grateful for the ensuing peace that has endured for longer than we have ever known.'

The man stopped walking and looked into her violet eyes.

'I must leave soon,' he said. 'My plan is to attack before they attack us. More Saxons are coming. They will come ashore close to the Roman shoreline. I will need my Sarmatians. Messengers have already gone out summoning them, but I fear many will not arrive in time even if they ride hard through the night, which I know they will.'

'Should I go to the Isle of St Brigid?' she asked. 'To Ynys Witrin, where my work is, where I will be safe?'

'I will speak to the merchantman, the captain of the corbita, and ask him to take you. It is where he goes next. He is reliable, and will take you first to Ynys Witrin and the care of Myrradin.'

Ganhumara felt a tinge of fear. 'Perhaps I should be with you this time,' she pleaded.

'No, that cannot be. It will be too dangerous. You will be safer on Ynys Witrin.'

She knew he was right. 'When will it be?'

'At first light tomorrow. That is when the merchant wishes to set sail. Afterwards, I do not know how long we may have. It may be we fight with what we have. But it will be the easier if I know you are safe.'

She knew the discussion was finished, so she turned and they resumed their walk. Having accepted what must be, she changed the topic of conversation. 'What of "The Light" in the sky? Have you seen any more?'

'Twice more, far off in the sky, but not shining directly at us. They seemed to be connected with the Land of Morwenna. Can it be true what she says?' he said.

The woman replied: 'Is it because she is a sorceress?' But she instantly felt foolish, for they were both Christians, no longer believing in black magic.

'Well,' the man replied, 'it must be some natural phenomenon. The "Ones Who Knew" accredited every action to such causes, just as Myrradin and his forefathers did. We are not far apart in our *beliefs, only in our gods. We believe a lightning strike is God's work. Myrradin ascribes it to gods rather than our one, single God.'*

'So what is causing "The Light"?' she asked again.

'I do not know. All I know is as you do. They arise in the Land of Morwenna, occur momentarily, and then are gone.'

'They leave me with headaches, and visions of things I know are not real,' the woman said.

The man did not answer for his mind had switched back to the coming battle; one that he knew he would not win. Not this time.

They accomplished the rest of the short walk in silence, each lost in their own thoughts, and soon reached the headland, sitting down in their usual resting place where the large rock protruded from the earth to make a natural seat. She snuggled closer as he placed his arm around her shoulder. It could have been a few moments. Or maybe it was longer, as they surveyed the horizon. Nothing was different. Maybe hours passed. Maybe not.

All they knew for certain was that dusk approached. The sun had gone, the cacophony of birds had grown silent. As Ganhumara and her warlord continued looking out at nothing in particular they saw the narrow, blue-white beam of dazzling light shine from above the far-off cliffs of Morwenna. It shone towards them, at them and then high up into the air, towards the southern part of the sky. As if a giant hand had grabbed it, the beam oscillated, glancing momentarily over the very spot where they were sitting. It was over in an instant.

The woman held her hands to her eyes, the man did likewise. There was an intense pain in their heads, but more so in the woman's. After what seemed an eon the pain lifted. The visions, too, released their hold, but by then it was almost dark.

'Holy Mother of Caesar,' the man gasped. 'I have seen a light brighter than a thousand suns.'

'Me too,' the woman cried, as her vision cleared. She looked up with tears streaming down her cheeks. 'Look!' she cried, pointing

with an index finger at a sky still illuminated by the sun's hidden rays. 'Look!' she cried again. 'Look! There! In the sky!'

The man followed her finger. High in the sky a dark speck was falling slowly from out of a rapidly expanding, but distant, cloud of smoke. From the centre of the smoke a number of black objects fell more rapidly, before splashing into the sea.

Suddenly, something white, like a canopy, blossomed out from above the largest of the objects, slowing its descent. They watched, fascinated, at something beyond their comprehension.

The sky was now clear except for this one falling object, swinging lazily backwards and forwards in pendulum fashion beneath its white canopy. Finally, they saw it splash down into the calm, opalescent, blue ocean. The canopy billowed out around it then slowly collapsed, as the object itself hit the water and then righted itself. There was a distant explosion followed by a dark puff of crimson smoke. The canopy separated from the charred and blackened object that so shortly before had swung beneath it.

All was quiet as the two distant observers continued looking on, as if mesmerised. Above the waterline, the object rose ten or fifteen feet to a sharp point. At the lower part, just above the sea surface, were what appeared to be windows covered in strange, transparent material that the watchers on land had never seen before. They shimmered in the failing light.

Above the window and towards the blackened nose were strange, white markings the watchers could not quite discern, for it was still too far off and the poor light made it difficult to read. Even if they had been able, they were not written in Latin, the only writing they could understand. It was written in English and said:

AURORA VI
High Altitude Escape Module.
United States Air Force
CXFM130508

As the two watched on from their cliff top vantage point, the object appeared to have an internal glow; a light deep inside. They thought they saw movement. Then slowly the seawater around the object began to change colour as a large patch of bright green fluorescent dye spread slowly across the sea, glowing with a ghoulish, greenish light as an automatic mechanism dumped its fluorescent payload to help air-sea rescuers who would never come to spot the object and winch its occupants to safety.

The mechanism was not only early, it was 1,500 years too early. With it a different story was about to be played out.

By now the couple had been joined by a crowd, each of whom was as incredulous and fearful as the next. The currents and wind were carrying the strange object to the east, towards the Land of Morwenna, towards the coast and a long strip of rocks.

Ar-tur felt a shiver run through the body of his wife, Ganhumara. He looked at her. He could see she was sobbing. He was puzzled.

'They have come for me,' she said simply, wiping the tears away. 'They have come for me,' she repeated, almost inaudibly above the sound of the ocean waves breaking against the rocks far below.

∞

President Sanchez had summoned his Chief of Staff, Cliff J. Wurzburg, for an unscheduled meeting in the Oval Office.

'Look, Cliff. This killing of one of our agents in San Francisco has me bugged. You tell me it was just some sort of unrelated shoot-out with a drug-head. But I've got a bad feeling about it.'

'So far there's no indication there was anything other than what the initial reports said it was. The guy was in the wrong place at the wrong time.'

'Yeah! Yeah! I hear what you say, but it doesn't cheer me up. We have limited time on DEEP EARTH.' He looked at his gold Rolex watch. Forty-five days. We get it wrong and several million people will be asking me some tough questions.'

'You don't …'

But the president cut him short. 'It's on my head whichever way we cut this. And I want to play safe when it comes to people's lives. I want you to arrange a meeting first thing with Nelson Loadhammer.'

'The Chairman of the Joint Chiefs, Mr President?' asked an incredulous Wurzburg.

'You got it in one, Cliff. Besides, how many Nelson Loadhammer's do we have round the place?' He paused a second or two before adding, 'Get it done first thing tomorrow. Whatever else I'm doing … move it. Do the same for Loadhammer. Do I make myself clear?'

'Yes, Mr President,' replied Wurzburg, knowing from the sound of the president's voice that there could be no changing his mind.

CHAPTER 10

It was early Tuesday morning with a weak sun shining from a pale blue December sky. The Anderson family had decided to extend their holiday in Cornwall to cover the forthcoming Christmas period only eight days away. The disused lighthouse at Pendragon Point had been converted into holiday accommodation two years earlier, and consisted of a low-rise block converted tastefully into two bedrooms, a kitchen and a large reception area. A wrought-iron spiral staircase led up to the former lamp room fifty feet above. Below it was another double bedroom with en suite bathroom.

The upper room had large windows cut into its thick walls offering spectacular views across the bay to the cliffs of Morwenstow, or out across the ocean 150 feet below. It was possible, too, to see the rocks immediately below, on which the ancient and rusting hulk of a small freighter still lay on its side. Waves constantly surged up and over its partially corroded flanks, swirled up its sides and superstructure before caressing its one remaining and crazily leaning funnel.

On the other side of the headland, and a short distance from the lighthouse, lay a grassy field used as a car park and guarded by a sentry-style car park attendant's wooden hut. The choice of the lighthouse for the Andersons' holiday had immediately met with everyone's approval. On that first morning spirits were high.

'The weather looks good,' said Jonathan from behind his copy of that morning's edition of the local tabloid, *The Cornish Guardian*.

The main story had caught his attention, with its headline:

EARTH TREMOR HITS CORNWALL

Then came the main story, pictures of startled people and a few lightly damaged buildings. It read:

An earthquake struck central Cornwall in the early hours of this morning causing hundreds of people to telephone Bodmin Police.

'The earthquake measured 3.5 on the Richter scale,' said Ken Houghton, of the British Geological Survey, based in Edinburgh.

Mr Houghton said the epicentre had been Bodmin Moor, with possibly affected property between 40 and 50 miles away. Reports were still coming in as we went to press from people living in Plymouth in the south, Bude and Tintagel in the north-east, and Truro in the west. Mr Houghton added that, "although it was a significant event for the UK it was very minor on a world scale.

"There are 50,000 such quakes a year throughout the world, but only two or three in Britain."

So far there have been few reports of injury or damage.

'Can you feed Ben for me, Jonathan? I'm running a bit behind.' Katrina was busy at the kitchen sink, hand-washing some items of clothing as the load was too small to use in the washing machine. She added, 'Dolores is having a shower and you're going out shortly to do what you've got to do, so can you feed him now, before you go?'

'Sure,' said Jonathan in reply. 'Just been reading about earthquakes in the UK. One's just happened only forty miles away, in Bodmin. Quite a rarity.'

Ben was sitting in his wheelchair next to the table, a bowl of breakfast cereal in front of him. Jonathan put his paper down and moved over to sit next to him.

As he began feeding him muesli, one spoonful at a time, Ben was animated but mostly silent, occasionally grunting his satisfaction. He looked happy.

Katrina shouted over to Jonathan, responding to an earlier comment he'd made about the weather. 'Yeah, it looks good,' Katrina answered in agreement. 'Let's hope it stays that way. Has there been a forecast on the radio?' Then, before waiting for a reply, she added, 'You still gonna do your thing with the bits of wire?'

'Sure,' Jonathan responded enthusiastically. 'I wanna get it over and done with soon as poss.' He fed Ben another mouthful. 'It won't take more than an hour, if that.'

'Suppose you find nothing?'

'Then that's a result. In science nothing is often as good as something. Either way, I'll worry about it when I get back. Then we can all go for some lunch at *The Bush Inn*, over in Morwenstow. Or *The Smugglers* down by the quay. You'll like either of them, though I prefer *The Bush*.'

'What about Dolores and Ben?'

'They can come too. We don't often go out as a family.'

'Okay. So be it,' said Katrina. 'You're right. It's been a long time.'

In little more than twenty minutes Jonathan was outside, pacing slowly and deliberately around the empty car park.

It was nothing more than a grassy field on the coastline close to the Devon border. Usually it was full, but only in the summer season as tourists came to admire the spectacular beauty of the cliffs and seascape which collectively had earned it the title of "A place of outstanding Natural Beauty" in the tourist guides. As Jonathan carried out his 'experiment' overhead came the shrieks and mournful cries of thousands of gulls, guillemots and cormorants, as they wheeled first one way then another.

To Jonathan's back, and ten miles away across the bay to the north-west, were the black antennas, white radomes, hangars and low-rise buildings signalling the presence of former wartime Coastal Command station of RAF Cleve but for the past ten years the satellite ground listening station of GCHQ Bude.

In each of Anderson's hands as he walked were two long pieces of stiff, galvanised wire, pointing directly ahead and parallel to the ground. Two hours before, they had been ordinary coat hangers. But after reading the website of *The British Society of Dowsers* the previous evening, Jonathan had refashioned them so they were now straight, with four inches of each end bent at right angles, allowing them to be gripped lightly so each could swing freely. His hope was that the two metal coat hangers would swing and cross each other of their own accord.

He was well aware that in the process of carrying out his experiment he could easily be mistaken for a water dowser, someone who believed that water, treasure trove, ancient buried artefacts and rare minerals could be found by causing stiff wires, a forked tree branch, or a pendulum bob to react suggesting something of value beneath the ground. In any event it was a long shot, but the prize was that if it worked he would know a far larger truth other than whether dowsing was fact or fiction. If the coat hangers twisted in his hands it could indicate a link between the effects of BRIGHTSTAR and what his revamped theory now suggested.

Anderson continued pacing. He was tracing a path first one way and then the other in a carefully pre-planned route that, when finished, would mean he had criss-crossed an area roughly the size of three football pitches. With dogged determination he persisted in his efforts while the two wire coat hangers remained oblivious to his efforts.

He was midway through his task with, he estimated, another half an hour before he would be done. He came to the end of the last leg of a route taking him first north and then south across the car park. He intended repeating it again, but going east to west until he'd covered the entire car park for the second time.

He knew the alignment of BRIGHTSTAR for the test. It had pointed harmlessly out to sea, ultimately focusing somewhere into the middle of sea area Fastnet, one of the thirty-two sea regions surrounding the British Isles. These are used as locations in providing weather forecasts to shipping, particularly fishing boats. Fastnet stood off the southwest Irish coast at the gateway to the North Atlantic.

Pendragon Point was in direct line of sight with the GCHQ ground satellite station. He'd noted, too, over the years he'd worked on the project, how curious it was that so many ley lines, those ancient lines of force said to cover the earth's surface, converged or radiated away from Morwenstow. Some tracked out to sea with one intersecting Pendragon Point. Another ran off to the east, intersecting with Glastonbury Tor, once a major religious centre for the whole country.

Anderson had just reached his start point again. The sun was now quite warm, with the breeze off the sea having dropped to a gentle whisper. His maths had indicated that there might be some causal link between his science and the old beliefs of the ancients. The maths was never wrong unless the theory itself was wrong. He was halfway along the latest leg when it happened. Initially he was shocked, even disbelieving. But finally it filled him with excitement.

The two pieces of wire he was holding had suddenly swung, firmly and unequivocally, to point towards the sea 200 yards away and in the direction of Pendragon Point lighthouse. Anderson twisted round to check what lay at his back, not that he needed to know.

It was GCHQ Bude and Jonathan Anderson felt a surge of satisfaction.

CHAPTER 11

The forty-two-year-old former detective sergeant with the grizzled face was more used to chasing scum with a crack problem than watching aeroplanes take off. His last street case had been a seemingly motiveless crime, involving a middle-aged man shot several times in the back in an underground car park belonging to Baltimore's University of Maryland. It seemed like an age ago, although in reality it was only six months since he'd applied for, and got, a transfer to the National Intelligence Directorate (NID), where he was now a field agent.

Following the move, he'd quickly appreciated he was in a different ball game, and was still slightly puzzled as to why it was him they'd chosen – an ordinary, hard-nosed street cop – instead of the usual college graduate. But no matter, he was glad of the change. Instead of chasing crackheads down mean streets littered with rubbish here he was, watching a bright red and blue-liveried Southwest Airlines Boeing 737-900 preparing for take-off; Flight 43 bound for Albuquerque, 1,100 miles away in the state of New Mexico.

It had stopped momentarily while awaiting final clearance from Air Traffic Control, but was now taxiing with increasing speed before lift-off. Finally, and with mild relief, he saw its nose tilt skywards and the main undercarriage wheels unstick from the tarmac of Oakland International Airport, one of San Francisco's three major airports.

He looked at his watch. 10.15 a.m. It was right on time. Although Flight 43 was merely one of 104 flights Southwest Airlines ran daily out of the city, for the NID cop this flight was special. It carried Professor Robert Pahlavi and it was his duty to see him safely on board so that, in three hours' time, he would be walking out through Albuquerque's Sunport passenger arrivals lounge, and into the arms of former CIA operative Tom Santos who, according to the Directorate's readily accessible personnel database, had been with 'the firm' since it's earliest days five years ago. His profile made him out to be a particular hard case, with training at Fort Bragg in his background, along with special forces and the intelligence community. The NID must be a cakewalk for him, the cop had thought afterwards, as he'd closed the file.

As Flight 43 departed, the former Baltimore policeman's job had come to an end, until he was told when, and on which flight, Pahlavi would be returning. That could be days away.

Before turning round and walking out of the departure lounge towards the car park he faithfully jotted down in his notebook the happenings for that day, December 20th. He'd scarcely given a thought to Special Agent Martinez, who'd been doing this self-same job only a few weeks earlier. 'Killed while on active duty', was all he knew about him. That, plus rumours he'd heard about being the victim of some probable accidental killing by a probable crackhead looking to fuel his next hit. Martinez had simply been in the wrong place at the wrong time he, along with everyone else, had concluded.

It was this conclusion that meant he had no need to be suspicious of the two men of obvious Middle East extraction who had also boarded the flight at Oakland, one of whom was seated in Economy, row C21, eight rows behind Robert Pahlavi's first class seat, or of the other casually dressed Middle Eastern man also seated in Economy, towards the rear of the aircraft in row F35.

∞

Three hours later, Southwest Airline's Flight 43 was beginning its descent into Albuquerque's Sunport International Airport. Touchdown would be in twenty-two minutes, at 1.30 p.m.

Professor Robert Pahlavi was not a good flyer. He was already nervous, and when he had seen sheet lightning zipping from cloud to cloud, thirty minutes out from Albuquerque, he'd felt physically sick. Underneath his jacket, and a source of constant but mild discomfort, was the thin, pocketed leather belt fitted tightly around his waist beneath his shirt, similar to those worn by sales reps employed in the jewellery trade. But instead of jewellery or expensive watches the pockets in Pahlavi's belt held two external, five-terabyte computer disc drives, each a mirror of the other.

They were small, waterproof and virtually indestructible in their dull grey titanium cases. Each one was stamped 'Property of the United States Government. If found return to Los Alamos National Laboratory', along with an identifying serial number and an unbreakable chain allowing them to be worn alternatively around the neck, like military dog tags or identity cards, if preferred.

It had been impressed on Pahlavi's team that new security measures at Los Alamos National Laboratory, where he was heading, together with the very nature of his classified work, made him a high security risk. They had told him that if he must ferry classified data between Berkeley and the Lab then every precaution against unforeseen security breaches must be taken. Even accidental loss caused by human error would carry a high and automatic, but so far unspecified, penalty.

On the empty seat next to him was one of his most treasured possessions. It was protected by a clear plastic cover and was an original November 8th, 1948, edition of *Time* magazine, featuring a head-and-shoulders cover portrait of Dr J. Robert Oppenheimer. It carried the cover line 'The Eternal Apprentice'. With it were old and faded information pamphlets of Trinity Site, provided care of the White Sands Missile Range, of which Trinity formed a part. At some point during this trip, Pahlavi was aiming to achieve a boyhood dream. He knew this flight could be his last en route to visiting Los Alamos, so it had to be now when he would visit the site of the world's first atomic detonation. In 1975, it had been added to the list of US National Historic Landmarks. As it was part of the largest active military range in America, visiting was only allowed on two days of the year. One was the first Saturday in April. The second was the fourth Saturday of December and, as luck would have it, this trip would coincide with December's date. It was too good to miss.

On the map it hadn't looked far from Los Alamos, maybe 200 miles. On American roads it would be a breeze to drive there and back in a day. No one would know if it was time spent on the project or not. Indeed, he was ahead of schedule with his work for DEEP EARTH. There would be time. There would be plenty of time. There was no one to object. His usual 'protection' bodyguard was not there to say 'yes' or 'no' to where he went or what he did. And as for his 'new' escort of Iranians, they had assured him that they would keep well in the background. There was nothing to fear, they had said.

'Do what you normally do, so that you do not arouse anyone's suspicions.' He had told them of his intention to visit Trinity. They'd seemed quite pleased with his intention. It had quite surprised him.

Almost since childhood, Pahlavi had devoured all the books about the Manhattan Project that he had been able to obtain in his native Iran. He glanced again at his old copy of *Time*. A wave of excitement swept over him. The 737 banked as it prepared to line up with the runway, still some miles away and 20,000 feet below as they descended.

Pahlavi's mind shifted to when he had been a young professor of geology at Tehran University; a bad time for anyone associated with the Shah. With help from the CIA, the Shah had ruled Iran for the best part of forty years, modernising it and bringing in many much needed social reforms. But some of those had angered the country's fundamentalist religious leaders, leading eventually to his overthrow. At that time, Pahlavi had specialised in earthquake prediction and vulcanology. As one of the most seismically active areas in the world, Iran had special need of his knowledge. Several major fault lines crossed ninety per cent of the country, resulting in an average of one major earthquake disaster a year.

Other countries other than his own frequently sought his acknowledged world-class expertise. His own suspicions that his country's leaders mistakenly believed he had once been sympathetic to the Shah's regime meant he was constantly on guard. He was always looking over his shoulder at shadows, always suspecting the motives of others. Ultimately, suspecting everyone of everything had almost led him to a nervous breakdown. When he'd been living in Iran he'd known there was only one cure: to leave the country. He had lectured on vulcanology in Britain and had developed close ties with Imperial College. It was why he had first gone to Britain and not the United States. Then had come the Berkeley job offer. Not only did it offer more pay he believed he would be safe from his countries agents, the same agents who had killed so many in such places as Paris. But in the event it had not been enough. Even in the US they had pursued him so that even now he was a willing tool fin their hands and all to protect his beloved son, still in Tehran, and with so much of his young life still left to live.

As the 737 approached Albuquerque airport, Pahlavi realised he had been a long time out of Iran. The result was that the last major Iranian earthquake he had been involved with personally had been in 2003, when the ancient city of Bam in the south-eastern part of the country had been flattened by an earthquake of 6.6 on the old 1930s-formulated Richter scale. It had been superseded in 1970 by the Moment Magnitude Scale (MMS) as the standard measurement of the world's geological community. Such a development was an irrelevant nicety to the 30,000 victims killed, 30,000 injured and 75,000 left homeless.

But it had been the smaller earthquake disaster at the city of Tabas, Eastern Iran in 1978, when 25,000 had been killed, that had lent intellectual impetus to his efforts in developing more sophisticated computer modelling techniques, as an aid to predicting when and where an earthquake might occur. One of his two PhDs had covered his work in this area. His second had come from the University of London's Imperial College, with his follow-up work at the University of California, Berkeley, which had led to his current work with the DoE.

At Tabas, he had sent teams out into the field before the earthquake had struck, recommending safety precautions that research suggested would save lives. Ironically, it had been the Shah personally who had supported his efforts most enthusiastically. Many lives had been saved, even though precautionary work had only been partially completed. It was the unintended by-product of a newspaper photograph, showing him shaking hands with a grateful Shah. It was from that very act all his personal troubles had subsequently flowed.

The essence of his modelling work lay in first recording all relevant earth movement data of the area under study. It was a tall order: the better the data, the better the forecast. To do it accurately enough to be meaningful meant enough manpower to collect the data, and then supercomputing resources for its interpretation. There was only one country on earth with those in abundance: the United States. The supercomputers were usually only available through a United States national laboratory, of which there were seventeen. Of these, only two had supercomputers powerful enough for his needs: Los Alamos in New Mexico, and Lawrence Livermore in California. He would have preferred to use the facilities at Lawrence Livermore as it was closer to where he lived and worked in San Francisco. But when they had asked him to work on DEEP EARTH, the Department of Energy had claimed their Lawrence Livermore facility was already overloaded with work resulting from its own National Ignition Facility fusion research programme.

So here he was, making what he hoped would be his last trip to Los Alamos to use their recently uprated ASC Roadrunner supercomputer. Its designer and manufacturer, IBM, claimed its raw processing power was equivalent to 100,000 fast laptops. In lay terms, it was the fastest, most powerful computer on the planet.

Pahlavi had never been told what DEEP EARTH was about, although he could guess. It had come as part of a project they were just completing where he and his twenty-two-man team had recently finished a geological re-evaluation of three of the DoE's existing or proposed nuclear waste sites. The first had been at the 600-square-mile Hanford site in Washington state. The second was at the Waste Isolation Pilot Plant (WIPP) at Carlsbad, New Mexico. Finally was the much-delayed Yucca Mountain nuclear waste depository in Nevada.

The objective, he'd been told, was to reaffirm the long-term geological stability of those areas, and by 'long-term' the DoE meant at least 250,000 years: the half-lives of the thousands of tonnes of radioactive materials either already stored, or expected to be stored some time soon.

Originally, WIPP was to be home to low-level waste, such as contaminated clothing and equipment. Then the requirement had changed to high-level waste from power station reactors and waste from the nuclear weapons industry. Despite the billions of federal dollars spent on its construction, it was still not fully operational, primarily because of concerns expressed by the government's own Environmental Protection Agency. The proposed Yucca Mountain Store had fared little better. Its purpose too was the storage of high-level waste, derived from spent nuclear power station fuel rods, US naval reactor vessels salvaged from decommissioned nuclear-powered ships, or plutonium no longer needed for nuclear bomb-making, and recovered as part of the Cold War peace dividend. But Yucca was within the State of Nevada jurisdiction, and that state had withdrawn its welcome mat, despite long association with the nuclear industry, which had seen over 1,000 nuclear tests, many atmospheric, at its Nevada Proving Ground. The result was to leave thousands of tonnes of highly toxic nuclear waste with nowhere to go.

This left only the Hanford site as an option, already classified as the most contaminated place on earth. Along with Oak Ridge in Tennessee, it had been the original plutonium and uranium production complex for the first atomic bomb and, since its construction in the early 1940s, had been in continuous production until its closure in 1991. The process had accumulated: 56 million gallons of radioactive waste stored in 177 leaking underground tanks; 2,300 tonnes of spent nuclear fuel, sitting in two leaking pools, a few hundred feet from the Columbia River; 120 square miles of contaminated ground water; and 25 tonnes of plutonium with nowhere to go but, in the meantime, so dangerous it was kept under armed guard. And by default, Hanford kept attracting more and more dangerous waste because there was nowhere else.

Robert Pahlavi knew all of this, just as he knew that for Hanford, time was running out, and that the Department of Energy's belated $230 billion clean-up campaign was going to be too little ... and far too late.

Although he had not been told anything more than he needed to know about DEEP EARTH, the rest had been easy to figure out. The federal government was going to try and contain the problem by burying it using either a single nuclear device or more. It was risky but, in the time Pahlavi believed was available, there was simply no other practical solution. The alternative was to have the waste covering the states of Washington and Oregon, maybe even the Canadian province of British Columbia. It would be so toxic that a major evacuation would be called for, on an unprecedented scale. Back in the war years of the 1940s, no one had considered these environmental problems, for there was a war to win.

Originally Hanford had been chosen for its remoteness and ready access to megawatts of hydroelectric power, from innumerable hydro-electric schemes along the Columbia River. No one had then thought of it as an earthquake zone. It was only much later when realisation had come that it was just a matter of time before the 'big one' would come; maybe today, maybe tomorrow ... maybe in 200,000 years. But, one day, it would happen. That much had always been certain. Almost daily proof, if any was needed, was given by the state's volcanically active Mount St Helens, in the Cascade Mountain range.

According to Pahlavi's admittedly early predictions, Mount St Helens' eruption in the 1980s had been merely a forerunner of what was coming. Deep underground fissures were already opening up beneath Hanford as the earth's tectonic plates beneath it shifted; a knock-on effect from the Pacific Ocean plates moving eastward. Normally, it was only a matter of a couple of centimetres a year, but now it was more and occurring at greater speed.

It was why there was a San Andreas fault line, running along much of America's West Coast. In 1905, it had devastated much of San Francisco, with minor damage again in 1989. Hanford was a time bomb.

Its leaking underground storage vessels were already contaminating human water supplies. Rising levels of radionucleides in the Columbia River was further proof. It was little wonder the region's sockeye salmon were becoming deformed or switching gender.

That was as much as Pahlavi knew for sure. The rest he could only guess. If DEEP EARTH did not work, then millions of Americans would be affected, maybe even die. There was worse. If his calculations at Los Alamos were incorrect, then DEEP EARTH might not prove a saviour, but could instead trigger the very disaster they were hoping to avoid. It was why his computer model was so critical.

Each time he flew out from Oakland, he had with him the latest results of his calculations and modifications, saved on to the five-terabyte, holographic storage-encrypted discs he carried, which also had the dual effect of acting as digital keys, which had to match the digital lock allowing him access to his authorised Roadrunner computer partition.

As it was the new updating, planned for this current visit, would run over several nights, ready for him to verify each following morning. Although Pahlavi could not be certain, his view was that although the margins of error for DEEP EARTH were going to be narrow, his best guess was that they would prove acceptable. The final go/no-go choice had to be the government's, maybe even involving the president. Los Alamos security had explained to him, time and again, that his work on DEEP EARTH was compartmentalised, so that those involved only knew what they needed to know. They had told him too that the Secretary of Energy was the only other person who had a computer override key. The problem would only be if both lost their keys. The chance of that was remote, virtually an impossibility, Pahlavi thought.

The Southwest Airlines 737 passed high over the north-eastern end of the two-million-acre Cibola National Forest. Immediately below were the 11,000-foot-high peaks of the Sandia mountain range. A few minutes later the aircraft was over the airport and, from his window seat, Pahlavi could clearly see far below two dark, cross-shaped strips marking two different sets of runways. They both stood out sharply from the light browns of the surrounding countryside, and the darker stain that was the sprawl of downtown Albuquerque.

The smaller of the two crosses marked the runways of the civilian airport, and intersected with the southern arm of the far larger and longer arm of the other cross marking the east-west arm of the 13,000-metre length of the world's longest runway. It was part of Kirtland air force base, the 53,000-acre home of the 377th Air Base Wing.

Named after Col. Roy S. Kirtland, pioneer of the US Air Corps, among the base's other 200 client organisations was the US Air Force's Weapons Laboratory, the US Navy's Weapons Evaluation Facility, and the development centre for military spin-off from NASA space research.

Hidden deep underground were recently constructed magazines housing 2,000 nuclear weapons. These were constantly being shuttled back and forth, either out to other military units for operational deployment, or for updating or destruction at the Department of Energy's site at Pantex, Amarillo, Texas, 200 miles due east.

Also just visible from the 737's passenger cabin windows, and lying adjacent to Kirtland, was the Sandia National Research Laboratory. Originally, it had been derived from the Los Alamos National Laboratory where it had once been part of its 'Z Division'. In 1979, it had become a national laboratory in its own right. Sandia's main responsibility lay with the update, maintenance, research and development of specialist but conventional high-explosive triggers, needed to initiate a nuclear chain reaction.

Together, Kirtland and Sandia employed 21,000 people, contributing an annual $2 billion to the local economy. It was all part of New Mexico's original Manhattan Project dividend, accounting for 42 per cent of a $7 billion annual budget allocated to the National Nuclear Security Administration, and for funding core weapons research, development and testing. It was a big number that, in real terms, far exceeded even that reached at the height of the Cold War two decades earlier.

Pahlavi heard the engine note change as the Boeing 737 continued to lose height. He had temporarily forgotten about his two extra shadows somewhere on the aircraft.

'It's nothing,' they'd told him. 'Go about your business as you usually do. It's just reconnaissance for us.' Apart from asking him where he would be staying and where he would be located, there had been nothing else. They had seemed most relaxed.

CHAPTER 12

The Iridium satellite phone belonging to field agent Tom Santos buzzed quietly. But it had an insistent manner. Aged sixty-three with short, grey-flecked black hair, Santos stood well over six feet three, weighed 224 pounds, and had been watching approaching Flight 43 for some time, just as he had done on innumerable previous occasions. No one had ever told him why he was riding 'escort', only that it was important and he was there to make sure no harm came to his charge. He was there to do a job as professionally as he could. It's what the feds paid him for. He was more proud of his role as a US federal marshal, and saw the state of New Mexico as 'his' territory. It was why he made it his business to know it like the back of his hand. Asked about the state's 'cowboy' past he would immediately warm to the subject, talking of the likes of the infamous William Bonney, aka Billy the Kid, Sheriff Pat Garrett, and of course the Lincoln County range wars of the 1880s.

The sat phone buzzed again. Santos fished it out of his leather jacket pocket.

'Yeah,' was all he said, after making sure he was out of earshot of anyone else who might be interested.

'Okay,' he said. He listened some more, then said 'Sure', followed by a single 'Okay', before he shut the cover and replaced the receiver back in his pocket. The message was from the NID and warned him to be on the lookout for two Middle Eastern looking gentlemen wanted in connection with an agent's murder in San Francisco. New information had come to light they had said along with descriptions. The message gave him brief details, along with the information that there were, so far, no photographs or drawings. Only they were suspected of being of "middle eastern" extraction.

He resumed looking out of the airport window, his left hand pushing his battered white Stetson hat further back on his head. Together with his well-scuffed, brown cowboy boots, Levi jeans and black and white check shirt, he was dressed in a style similar to the myriad other native New Mexicans waiting in the arrivals area.

Two days before he'd received his regular call, giving date and time for his regular human collection, along with further information saying this would be the last time. The job was over. He should expect his 'assignment' to come to an end.

Santos was an old and experienced hand who, over a lifetime in similar roles, had evolved his own particular way of seeing through each assignment. It was how he'd stayed alive, he'd sometimes said to anyone with the authority and balls to ask. Experience had long taught him to check, check and double-check, then check again. Check as if your life depended on it, for one day it might, just when you're relaxing, just when your guard was down and least expecting trouble. Sure as shit that's when you got it. It's what they taught you in special forces. There he'd learned to listen and, once or twice, had been grateful he had.

Just as he did on each and every assignment, he'd checked his .40-calibre Glock 22 service pistol. He'd chosen that calibre of weapon because the mid-size Glock 23 didn't have the extra two rounds, and he'd always maintained that, one day, you just might need 'em. Just as always, checking and cleaning your weapon might prevent jamming. Otherwise, just when you didn't it would. It was the way he'd been taught, so that now it was his custom and practice. He continued to do it even though the Glock fired just as well wet or dry. Even if it had been immersed in the family dishwasher it would still work they'd told him. It hadn't changed his practice.

The phone call had come from HQ, first advising him of what he already knew. 'Take Pahlavi to the usual places and anywhere else he wants to go.' Then, 'Escort him back to Albuquerque airport.' The usual applied. If there was to be a fuck-up, then it wouldn't be on his watch.

Then came the new assignment. 'In eight days' time, Energy Secretary Hoop Toberman will be arriving for further talks with state officials about the future of the Waste Isolation Pilot Plant and its role in handling long-term nuclear waste.' Santos had snorted when he'd heard it. WIPP had already cost $19 billion and there'd been over 10,000 transuranic waste shipments so far over the past ten years. There was sure gonna be a whole lot more over the next twenty to thirty-five years during which WIPP was expected to be operating. Problem was, as time went on the state was getting more and more touchy about its role as the nation's nuclear dustbin. He'd heard the feds wanted to expand WIPP, while the state legislature wanted to close it down.

'Afterwards, he's flying directly to the UK,' the voice on his phone had intoned, 'for talks with his opposite numbers, on something to do with a clean-up programme involving a British company contracted to the US Department of Energy.

'Severe time constraints mean the Energy Secretary will be using the Department of Defence's top secret AURORA flight system, for which Santos is the only NID agent cleared for its use because of his former CIA connections. Your knowledge of it could be useful in an emergency. So', the voice continued, 'you are to accompany the Secretary to the UK and back again. It should be a short, round trip. Flight lift-off from Kirtland is at 1900 hours. A six-hour time difference when added to the three-hour flight time, including in-flight refuelling, means it will be around 0400 hours local time when AURORA touches down at RAF Fairford, its UK landing point.' The message ended.

AURORA was intriguing stuff. Santos knew she was a leftover from a discontinued CIA 'black' project and later transferred to the air force. Originally designed as a suborbital space vehicle flying at Mach 6 at over 200,000 feet, she could reach anywhere on earth in three hours, excluding refuelling, provided there were vital facilities necessary to make her ready for the return trip. It usually meant pre-positioning a C130J transportable pod, flown out the previous day. Kirtland air force base, along with its lengthy runway, had all the necessary facilities, which was why it was the secret hub of all worldwide AURORA flight operations.

As Santos returned to watching the approaching 737, he remembered how Pahlavi had mentioned a few times that he'd always wanted to visit Trinity Site. He'd wanted to see some of the old stuff too, like Oppenheimer's old house, still standing in Los Alamos. 'Well,' he thought. 'It had better be now or never. Tomorrow will be too late.'

Passengers were coming off Flight 43 and entering the arrivals lounge. Santos in one glance took in that there were over 100 people like himself, mostly male, waiting to greet someone. He moved closer to the passenger exit point so that Pahlavi would be more likely to see him.

The weather outside the passenger terminal was clear, ground temperature 46 degrees, with a three-mile-an-hour light, westerly breeze.

Although Albuquerque was not the state capital, the city had grown considerably since the Second World War, thanks largely to defence work. Additionally, it was home to some 300 federal agencies and, along the way, had successfully fostered a burgeoning tourist industry.

Situated in the middle of the Rio Grande rift valley, connecting Colorado in the north with the Mexican state of

Chihuahua 300 miles to the south, the city was surrounded by the reds, whites and browns of the desert and the greens of pinon pines and cottonwood trees, all linked by a spectacular backdrop of mountains, set against permanently blue skies.

It had evolved into a spacious and well-planned modern city of 500,000 people, cut in half by the wide Rio Grande River and a mixture of high and low-rise buildings, where the old jostled harmoniously with the new.

The addition, in 1966, of the Sandia Peak ski and tramway, the world's longest, had also boosted its increasing popularity as a winter sports centre. And in the autumn its location made it a hot air ballooning centre, where the annual festival always attracted 500 colourful entrants from over 50 countries.

Pahlavi spotted Santos almost immediately as he came through the crowded arrivals section.

'Hi there, Robert! Good to see ya again,' Santos shouted. 'Glad ya could make it. Lotta things we gotta talk about.'

They shook hands and walked out of the airport towards the car park and the late morning sunshine. Pahlavi noted and readily appreciated the crispness of the air. With the sun strong on his face it reminded him of his native Iran. In that moment, it felt good to be alive, briefly forgetting again the new shadows he'd recently acquired.

Farhad and Kouros had already left the terminal building and were among the last passengers to leave Flight 43. They'd taken care not to be seen together. For all anyone knew they were just two youthful passengers who, despite their obvious Middle East origins, could have been anybody going about their lawful business. Only Tom Santos had noted them.

CHAPTER 13

'We're waiting for the final computations from our Iranian friend,' said Chief of Staff Cliff Wurzburg. 'He's at Los Alamos even now, carrying them out.' The president was seated with his back to the window that looked out on to the Rose Garden. His face expressed concern. Wurzburg noted it. 'I know it looks tight,' Wurzburg continued, 'but so far as we can see, DEEP EARTH is right on schedule. So we need to keep relaxed.'

'It's December 22nd. We've got 40 days left,' the president replied. 'And I've got the governor of Washington state here at eleven. I've got to tell her about the emergency exercises we're planning for her state. Since New Orleans and Hurricane Katrina we run 'em all the time in some state somewhere, so it shouldn't come as any surprise. The last big one they had was three years ago. She might think something's up because I'm involved. But she should be used to the attention, given the level of earthquake and volcanic activity they get.'

'Nancy Roberts is a smart governor, Mr President,' said Wurzburg. I know her and the state pretty well. 'State capital is Olympia with a population of 45,000 all living along the shoreline of Puget Sound. Nancy's used to emergencies, and so is Olympia. Had a few earthquakes: '49 and '65, the last one being in 2001 with the Nisqually 'quake. And at this time of year it'll be wetter than the Everglades. If there's one thing it does in Olympia, it's rain and rain.'

The president appeared not to be listening and merely said glumly, 'We've gotta tell her the next exercise her state'll be having will be a big one and'll go on for months. She won't like it.'

∞

There were only seconds to go. The project controller had been passed authority to commence BRIGHTSTAR's ignition sequence. There was silence, an air of expectation from the people in the control room who were sitting close by. At GCHQ Morwenstow it was five o'clock in the early evening. Outside it was already dark and beginning to rain a light rain. On the surface, among the hangers and helicopters, there seemed to be fewer soldiers than normal going about their duties.

'Five,' said the measured voice of the project controller.

'... Four ...

... Three ...

... Two ...

... One!'

His finger pressed a small, red button. 'We have ignition sequence,' he said expectantly. There was a prolonged silence. Nothing extraordinary seemed to have happened. All the gauges and dials still showed nominal. But for those present they knew the test, the fourth of its kind, had been a complete success. Gradually they had been increasing BRIGHTSTAR's power setting, with the last being within the 'intermediate' range. The next test would be at full power and with a target. So far it had all gone exactly according to plan. Even so, the relief of those present was palpable.

Outside nothing seemed to have changed. It would have taken the most observant of observers to have detected the thin, blue light that shot for an instant high into the air and lasted in actuality for thousandths of a second.

∞

With Pahlavi sitting beside him, Santos drove the Jeep out of the airport car park and into slow-moving traffic heading west down Sunport Boulevard. Within a few minutes he turned north for Los Alamos. He'd already noted the grey Pontiac truck that could be following them discreetly three of four vehicles back. It could just be coincidence that it had made all the same turns they had; too early to judge. But if they were hostile then they were rookies, Santos thought; too eager and too close. They hadn't had much time or back-up either, otherwise they would have planted a homing device. He could be paranoid after too many years in the job. More than likely the Pontiac belonged to an innocent rancher going about his business. He hoped so. Otherwise he could maybe look forward to a little excitement.

'I'm so happy to be meeting with you again,' Pahlavi said by way of conversation. 'I have my work to do as always, but I'm hopeful there will be some small amount of time left for other things.'

'Oh, yeah?' said Santos. 'I hope you don't mean women!'

'Oh no, nothing like that.' Pahlavi had coloured up at the mention of 'women'. He laughed a nervous laugh. 'No, I'm hoping to visit Trinity Site in what I believe will be a spare day at the end of my work.'

'It's not too far. Two hundred miles,' reaffirmed Santos. 'Of course, I gotta come with you just in case.' Santos overtook a slower-moving car. 'Look on me as your fairy godmother. You know the drill. Where you go, I go, so long as you're on government business.' It was his turn to laugh, but it was an easy laugh as he felt the gentle pressure of his shoulder holster underneath his left armpit, as he swore at the driver he was overtaking.

Pahlavi laughed too. 'You know, this country has been very good to my family and myself. It's an honour for me to be doing the work I do for your government.'

Santos overtook two more cars. 'Tell you what I could do. I could arrange for you to stay at the old Fuller Lodge, rather than your usual Comfort Inn on Trinity Drive. Even though it's been refurbed, it's still a dump. Fuller Lodge is right in the centre of Los Alamos. Could be more your style, seeing as how it'll be your last visit here. I'm sure you know that before the war, it was where the "Boys School" used to be, which they used as the base for the original labs. When the military kicked 'em all out it was where Oppenheimer's scientists had their fun and games after a hard day's night designing the Bomb.'

The words 'Trinity Site' set Pahlavi's nerves tingling with almost boyish excitement. 'You could do it?' he heard himself blurt out. 'You could do it now?'

'Sure. Just give the word and I'll fix it on my mobile. I'll phone ahead. I've got both the Comfort Inn and Fuller Lodge on my speed dial. It's no big deal.'

Pahlavi thought about it for a few more seconds before acquiescing. Santos was as good as his word. Within a few minutes the change had been effected and his existing reservation cancelled. As Santos put his phone away he said, 'You know, they used Santa Fe as the postal address for Los Alamos when they were first building the labs. It wasn't secret like everyone thought. Nope. Los Alamos was jus' too small a place for the map makers and the US postal service, so all the mail went to PO Box, 1663, Santa Fe, New Mexico. Until all the scientists arrived, there was nothin' there. It was just a small boys' school ... and the fact that Oppenheimer liked that part of New Mexico. It's where he went horse-riding as a boy.'

Around them the landscape was slowly becoming redder, with huge, weather-eroded boulders littering the countryside. Occasionally there was a thin covering of snow, and where the road sliced through the earth it exposed pumice, a leftover from the huge volcanic blast that had formed the state over a million years before.

'All the reasons Oppenheimer gave about Los Alamos being remote, making it the ideal place for siting his new labs, was so much bullshit. It was the horse-riding, his sense of humour and history too. It was here where one of the world's biggest volcanic eruptions happened over in the Jemez mountain range. It blew 250 cubic miles of rock into space. It must've been some bang, I can tell ya. It covered the whole state with six feet of pumice. No kid.'

Santos concentrated on the road ahead before continuing. 'And was it coincidence that the Anasazi Indians, the Ancient Ones, the "Ones Who Knew", the old tribe of Pueblo native Americans, had lived there? I don't think so. Oppenheimer was too smart. He saw himself treading in their ancient footsteps. At least, that's my theory.'

Interstate 25 eventually carried them around Santa Fe and on to Highways 502 and 4, before taking them through the small town of White Rock with its identical, sanitised homes and neat, suburban layout. Eventually they passed a blue-signposted narrow road, leading off into a scrub-covered part of the mesa. The blue-signpost said simply: 'Technical Area 39'. There were more, similar signs and roads, with similar signposts announcing 'Meson Beam Facility 13', or 'Explosives Area 21'.

The outskirts of Los Alamos were evidenced by the now-empty watchtowers, high chain-link fencing and satellite communications dishes. It was a town too, that although its glory days might lie behind it, still had an annual federal budget in excess of $3 billion.

Its wartime nuclear past was honoured with street names such as Oppenheimer Avenue or Trinity Drive, mixed with more prosaic names such as Cerrillio Drive, with hotels and fast food outlets of Taco Bell, Denny's, Burger King, McDonalds and a Holiday Inn Express hotel. A quarter of the city's 18,000 population were said to have PhDs.

They drove along Trinity Drive, past the Bradbury Museum and the library, round by Ashley Pond, from which the first reactor had drawn its cooling water, now replaced by modern sculptures and fountains and all enclosed by a green park. A left turn and they were drawing up at the brown, semi-timbered building that had been the historic Fuller Lodge.

As their car came to a halt outside it Santos volunteered additional information: 'This road was called Bath Tub Row by everyone working in the red quagmire that was the original camp.' With his arm Santos indicated the single-storey brown wooden buildings lining the rest of the short road. 'Those buildings,' he indicated, 'used to house the hotshots in the old days, and were the only houses with hot water for a bath!' He smiled wryly at Pahlavi. 'So now you know how it got its name!

'Next door was Oppenheimer's old house.'

With that he got back into the driver's seat and watched his charge walk slowly towards the entrance of Fuller Lodge. He shouted after him as he went: 'So how long have we got before heading out towards Trinity Site?'

Pahlavi stopped and turned in his direction. 'I have six full days at the laboratory. We go to Trinity on the seventh, before I go back.'

'Okay, Prof. I'll do the usual escorting to the gate of the lab tomorrow, from where they become responsible for your safety. Then, in seven days' – he checked his watch – 'that'll be Saturday, December 29th, we'll both head off to Trinity.'

CHAPTER 14

Sir Freddie Stirling MP, PC, KVCO looked idly through his helicopter window at the low cloud layer sliding past 500 feet below. He was well on his way to visiting what the British Prime Minister had referred to as the 'Crown Jewels' of UK defence policy. 'So secret,' the PM had added, 'that only a handful of people know of its existence.'

He had been referring to BRIGHTSTAR, and Sir Freddie was going to spend most of the day looking at his new charge. He was going to be there too, into the small hours of the morning, when a 'low power' test would be conducted. He only wished it could have been a few days later, when the planned 'full power' test was scheduled to take place against dummy targets. But it was not to be. He checked his watch. It was Christmas Eve, not the best time to be visiting the BRIGHTSTAR site but that's what his schedule had demanded.

He was in his mid-sixties with lengthy, brushed-back white hair, grey pin-striped suit and tortoise-shell glasses, giving him the appearance of having beady eyes. Years ago, he had been a major in the Royal Tank Regiment, and could still sound like it. He still wasn't too sure why he'd got this new job. Maybe it was because he'd once been Shadow Defence Secretary while in opposition. Maybe it had been his long association with the defence industry. Or maybe it had been because of his chairmanship of the Security and Intelligence Select Committee. One rarely knew how these things came about. At sixty-six, he was too old to really care.

The British Army Westland Augusta AW159 Super Lynx helicopter was cruising at 180 knots. Sir Freddie was wearing ear defenders to minimise the whine from its twin 1,362 horsepower Rolls Royce engines. Papers were spread on the empty seat beside him. Behind were the pilot and navigator. In front was the six-man passenger cabin, plush compared with the usually spartan conditions of most army helicopters. Having its use was a perk of his newish job as cabinet secretary to the Joint Intelligence Committee, reporting directly to the Prime Minister. He was the 'keeper' of state secrets and deniable actions. He was on a 'familiarisation visit', as the PM had called it.

'It's our Crown Jewels,' the PM had said. 'Time you were brought in so that if we ever have a problem you'll know what we're talking about.' With that it had been arranged, and here he was, at a time of great importance for 'The Project', he had been told.

He glanced at his watch. In ten minutes they would be landing at what was still called GCHQ Bude, but was actually sited closer to the village of Morwenstow than to the town of Bude. By ministerial car the 200-mile journey would have taken hours. By helicopter, it was 75 minutes.

Shortly he was going to be shown Britain's most secret defence project. It was on a par with the first atomic bomb, the PM had whispered, after a Cabinet meeting. 'So secret even the Americans don't know. You know how leaky they can be. Telling them is the same as advertising it in *The Washington Post*.'

Now he was going to be admitted into that inner circle, with those few who did know. Moreover, that evening there was going to be a 'shakedown test'. This had puzzled him initially. Nuclear tests did not normally take place within the shores of the UK. So what were they testing? His puzzlement had only cleared after reading the briefing paper, handed to him just before take-off. Before reading them he had mused over the development of the first atomic bomb sixty years before, and the secrecy surrounding it.

He'd read that when Vice President Harry S. Truman had taken over as president, on Roosevelt's death in 1945, he'd had no idea of the Manhattan Project's existence. He did not know about the huge plutonium-manufacturing sites at Hanford, nor of the equally huge uranium gaseous diffusion plant at Oakridge, Tennessee. He'd known nothing of the Los Alamos weapons laboratory, nor of the first nuclear test at Alamogordo, later explained away as an Army ammunitions dump explosion.

The briefing papers lying on the seat by Sir Freddie's side had, in part, provided answers as to why he was making the journey. They had been handed to him with the provisos not to read them until take-off, and that they would be collected on landing.

The briefing was short, written on six sheets of closely typed yellow paper headed 'BRIGHTSTAR'. It gave an overview of a new and innovative missile defence system, what to expect, and what lay behind it. Even for Stirling, a man who had known and held many state secrets, the contents took his breath away.

Although similar to the long-since-abandoned STAR WARS programme announced by fortieth US president Ronald Reagan in the mid-eighties, it showed how science had moved on. The science profoundly impressed him, but left him with a feeling in the back of his mind of 'will it work?' There was an undoubted need for it. No longer did the West have the comparative safety of just one evil superpower to contend with. Instead, there was a flurry of smaller nations, ranging from North Korea to a take-your-pick choice in Asia or the Middle East. Under the guise of developing peaceful nuclear energy programmes, many were also developing nuclear weapons, along with their means of delivery.

The briefing paper had two long, general introductory paragraphs on new 'energy weapons'. Emphasis had been given to 'Electro Magnetic Pulse' weapons that could interfere with electronic devices that now formed the basis for the development of 'Directed Energy' weapons.

They also included facts relating to the 1978 Bell Island mystery, when an energy bolt, equivalent to an estimated 10-megaton weapon, had affected this small island off the coast of Newfoundland. Although there had never been a satisfactory explanation, mention was made of it being coincidentally close to the US Brookhaven National Laboratory of High Energy Physics on Long Island.

The briefing discussed the development of EMP weapons, possibly for an 'anti-missile shield'. The problem had always been how to produce such a pulse without a nuclear explosion, and focus that pulse into a beam that could be directed and reproduced at will.

The British, the briefing said, were now perfecting such a weapon. It had the advantage of being virtually harmless to people, just as magnetic fields were harmless, but highly damaging to electronic components used in weapon systems, such as intercontinental ballistic missiles. And, owing to weight considerations, such missiles could not use extra shielding as a countermeasure. So such a weapon, the paper contended, should be 'future-proof'.

If the current range of tests proved successful, the paper continued, then it would radically reduce the number of sites needed to provide a complete missile shield for the whole of the British Isles.

The theory was that maybe only one or two 'power-generating' sites might be needed, along with associated, but essentially passive, 'reflecting stations'. Limited development of this had already been undertaken between Morwenstow and a 'reflecting' station on the Mendip Hills, close to Glastonbury, seventy-five miles to the southeast.

The helicopter's engine note changed as it began to sink towards its landing site at GCHQ Bude. Within five minutes, Stirling found himself automatically ducking the still-rotating helicopter blades and running towards a small group of people whom he presumed were his welcoming committee. He reached the group and straightened up to be greeted.

He noted that it was pleasantly warm for the time of year, despite a blustery wind sweeping in from the sea. The light was fading. Quickly his eyes surveyed all around. Nothing of significance escaped his attention. He noted the cliffs on the far side of the bay and the individual and bright, twinkling lights that seemed to be at the junction of sea and land; the cluster of different-sized white radomes, supposedly to cover radar antennae. He made a mental note that they wouldn't fool anyone who knew what they were talking about. Modern radars, of the type allegedly used here, no longer needed such protection, and had been phased out years ago, to his certain knowledge. He took in the trucks and the myriad of personnel still scurrying around. He sensed rather than saw their collective urgency.

'Sorry your visit here will be so brief, but even so, we hope to make it quite interesting.' The voice was that of a young woman dressed in an army uniform. He shook her offered hand, but cut immediately to the chase.

'What are the lights for over the bay?' he asked.

'Not sure,' she replied instantly. 'One of our beach teams became involved a short while ago.'

'One of your beach teams?' he queried.

'Yes. After each test, we take readings from the surrounding area.' She pointed a finger to the far side of the bay, to where he could make out the shape of a lighthouse.

'You see the old lighthouse? It's disused and empty now, replaced by a fully automatic light five miles out on Puffin Rock, which should switch on in a moment or two. It's possible that when we aim a beam out to sea, we might just skim the top part of the old lighthouse at our minimum elevation, not that it would harm anyone. Shouldn't be a problem. The beam is completely harmless, just as a magnetic field is harmless. It's only on for a short fraction of a second anyway.'

She caught the imperceptible look of surprise on his face. Stirling knew she had seen it, and silently cursed himself for allowing his poker player's face to desert him.

'We did another test run last night,' the girl said. 'Forty per cent of full power. We heard some dead seabirds had been washed ashore, so we had a couple of teams out this morning gathering them up for examination. Of all things, too, some archaeologists from Exeter have come across what they think is an unexploded bomb, or sea mine.' She smiled. 'That's livened things up a bit.'

She turned to walk towards the buildings 200 yards or so distant. 'Our beam only takes an instant, just an instant,' she repeated. 'We're quite proud of it, really.'

Perhaps it was then she realised her listener might not be fully conversant with the technology or anything else, for she stopped in her tracks and turned to look into his face before saying: 'Let's explain more fully once we're inside, and show you the full extent of our device. I hope you'll be impressed. There's a lot of it.'

He looked after her, to where the entrance lay. A figure was walking the short distance towards them, which he recognised as the stout figure of Sir Jack Geisner, boss of British Atomic Weapons Management. Sir Jack strode towards him offering an outstretched hand.

'Sorry I wasn't here to greet you, Sir Freddie. One or two last-minute problems. Nothing serious.'

The two began the short walk back towards the building. Sir Jack kept talking. 'You'll be aware, Sir Freddie, that BRIGHTSTAR is ahead of schedule. Subsequent to our Head of Laser Fusion leaving us, we thought we could and should advance our programme. This is exactly what we've done, and already we've begun the roll-out phase a few nights ago, with some tests at low power settings ...' Sir Jack opened the door and the two passed inside.

∞

At her present speed of thirty-two knots it would be another six hours before HMS Daring, a Royal Navy Type 45 Horizon Class destroyer, arrived at the MoD's gunnery and bombing range fifty miles west off Land's End in sea area Fastnet. In truth it lay more towards southwest Ireland than mainland Britain. Daring would be 'on station' for the next few days, with the 'test' itself being on or around 'the twenty-eighth'. 'Wait until further orders,' the Admiralty had told her captain.

The sea was calm, with only a slight swell to impede Daring's progress northward along the Irish Channel from Liverpool, where she had been paying a goodwill visit. Although not yet fully equipped with her latest planned technology, she was still a formidable stealthy weapons platform, as the Admiralty called its seventy-nine fighting ships.

In the dim red glow of the area defence operations room, the Principal Warfare Officer stared down at a clear-topped table. He was a tall and athletic thirty-two-year-old with sandy hair and a freckled complexion. Dressed in white anti-flash clothing, over which he wore his dark navy trousers and sweater displaying rank and name, only his eyes were visible through the slit in his anti-flash headgear. An integral part of his uniform was a communications headset through which he was receiving and giving orders. His concentration was absorbed by a bright horizontal screen of orange, forming a circle of light three feet in diameter, on which countless almost-white dots moved backwards and forwards.

The Welsh coast, the northern flank of Cornwall and the eastern coastline of Southern Ireland were displayed in outline. Between the Cornish and Welsh coast lay the Bristol Channel. Superimposed over this outline were a series of moving bright dots, each of which had a square of light with a number allocated to it. Each of the dots continued moving across the screen like two-dimensional fireflies.

Occasionally, the Principal Warfare Officer would move a cursor over one of the moving dots. Instantly, it would expand and the allocated box would reveal further details, such as height, speed, altitude and heading. Most were high-flying civilian passenger planes or RAF aircraft, whose Identification Friend or Foe (IFF) automatic transponders squawked back their identification number, corresponding to a specific number and moving box on the orange display.

At sea level, and picked out in a different colour, were other more slowly moving objects. These were identified only by type, such as fishing boats, container ships, passenger liners or yachts. Yet another colour depicted subsurface activity. One such colour corresponded to a Royal Navy, Trafalgar Class attack nuclear submarine making its way down the Irish Sea from its Faslane base on the Scottish west coast, heading for a rendezvous with Daring as part of the planned, joint exercise.

No one spoke among the fifteen other naval ratings in the operations room as each was absorbed by their own glowing monitor, each linked into the Principal Anti-Air Missile System (PAAMS). The Daring class destroyers were slowly replacing ageing Type 42 veterans of the Falklands campaign, which had been shown wanting in the modern missile age.

Daring was cutting-edge technology of which the navy was rightly proud. Once primed, the ship's defence systems were entirely automatic for, in modern warfare, things happened too quickly for human beings to respond. Fractions of a second could mean the difference between life and death for the whole ship, its company and any other ship it was its duty to guard.

The Principal Warfare Officer, together with the other crew members, had received a special briefing a few days before, on the nature of the forthcoming test.

'Mostly it will be routine,' the captain had told them. 'All I can add is that it is to test a new, land-based weapons system, where we will be acting the part of the "target". We've done it before so nothing new. Everyone will need to be on their highest guard. That is as much as I can say. The scientists we have on board are here to monitor our performance.' With those words the captain had dismissed his men.

The Principal Warfare Officer was sufficiently experienced to know there was only one real way of testing out the ship's defensive armament for sure, and that was by letting off a nuclear airburst. The resultant electromagnetic pulse, the EMP, frazzled the electronics even if housed in hardened casing. There was no likelihood of that. Instead, they would have to make do with decoys, aluminium chaff and similar toys. He could understand that. It was the rest that disturbed him.

What were the 'special sensors' fitted hurriedly to certain parts of the hull just before they had departed from Bristol? And who were the new men who had joined the ship, and what precisely would they be doing? Unaccountably, he had a bad feeling about it all. Mentally, he shrugged his shoulders. He would just have to get on with the job, see what happened, then scold himself afterwards for being so suspicious over nothing.

Strangely, for this exercise, there would be no live missile firings. For that they would need the Aberporth missile range off the south-west coast of South Wales. They had already been tested during sea trials months before. And the test they were practising for was, unusually, scheduled for 0345 hours, designed, the captain had said, to check vulnerability to enemy electronic countermeasures.

The Principal Warfare Officer's screen was pivotal to the defence of the whole ship. It integrated information from a number of different battle stations via a distributed SAMSON area defence computer system.

SAMSON gave a layered, defence-in-depth capability through its state-of-the-art technology. This controlled every element of the ship's weapons systems, and was designed to give a totally automatic response to any given threat. The offensive and defensive weaponry it controlled ranged from the forward-mounted rapid-fire 4.5-inch gun pod with its thirty-mile range, to active homing, long-range surface-to-air Aerospatiale Aster 15 missiles. There was a new, long-range system on board too, with its British Aerospace-developed 'Odin' that could survey for threats up to a height of over 100,000 feet. At the other end of the scale and for last-ditch defence, there was the point-to-point Aster 30 vertical launch missile system. This was further backed up by forward and aft-mounted Vulcan rotary six-barrelled cannons, each firing 3,000 rounds of depleted uranium shells a minute.

The SAMSON area defence radar system would estimate the time of arrival of each dot if it changed course to pose a threat, by moving in the direction of the ship. It would automatically allocate what weapon systems to use, and in which order, so that it offered an in-depth, layered defence response. Only computers offered the reaction time necessary to guarantee safety. In wartime, it would take less than a second from detection of an enemy target to response. In sea trials, it had proved very effective even against potential subsea threats.

The entire ship's company knew the coming test would be in sea area Fastnet. They guessed too that it would not last long. The Principal Warfare Officer knew that at 0345 hours, and on orders from the captain, he and a fellow Principal Weapons Officer would simultaneously turn two keys, setting the ship's defence systems from manual to fully automatic. To ensure safety, because no live firings were involved, only one key would be turned fully. The other would be turned halfway. This armed the computer system, but left the full battle and missile launch sequence inert.

∞

'Where is he? Where's he gone?' asked Kouros nervously.

'Don't worry,' answered Farhad calmly.

'He's tricked us. He was supposed to be here. Where is he?' Kouros retorted.

Farhad remained calm. 'I don't think he's tricked us,' he said in Farsi, one of the languages of his native Iran. 'If I'm any judge of men, he's not the type. He's not brave. He doesn't do, and won't do, brave things. All he wants is to continue his life, his very good life, as a Berkeley professor, helping out his native country's greatest enemy. And perhaps preserving his long-dead son's life.' He chuckled as he said it. 'He'll contact us. He can't be far. Be patient.'

Although they were in a non-smoking guest bedroom, located on the second floor of the Comfort Hotel, Kouros was nonchalantly smoking a cigarette, seemingly oblivious to a ceiling-mounted smoke detector. He then looked nervously out of the window, scanning up and down Trinity Drive for anything that might alarm him, such as a state trooper driving his cruiser into the hotel's parking lot. There was nothing.

'You must speak English Farhad, at all times,' Kouros said. Farhad retorted. 'Subsequent to your precipitate action with the American back in Sausalito, the police may already be looking for us.'

'No, Farhad. I left nothing for them to go on. I was careful,' Kouros said.

'Then why are you looking nervously out of the window?' Farhad blew a stream of smoke out of his mouth.

'One can never be certain of these things,' Kouros answered.

'No, that is true.' Farhad inhaled more smoke. There was a silence neither wanted to fill. At last it was Farhad. 'He will phone. Trust me. He's the sort. And if he doesn't, we know where he'll be. Either way, he'll help us with the plan. In his eyes he has no choice.'

'Supposing he tries to trick us?' asked Kouros.

'He isn't the sort,' answered Farhad. 'But he's nervous, and it was because of this we came up with our plan. It has simplicity. That's our best defence.'

'Why did we not get him to alter the data on the discs and ruin the Americans plans?'

'Because his nervousness in doing it would have given us away. Besides, what he could have done would have been limited. The Americans aren't fools. Their security, he's told us, included self-checking measures at Los Alamos. Anything outside of strict limits would have been questioned.'

Kouros remained silent, distracted momentarily as he continued looking out of the window. He could see a small boy he estimated to be aged five or six, who'd kicked his ball against the side of their otherwise anonymous grey Pontiac truck. The ball had rolled underneath the chassis and now the boy was attempting to retrieve it when his mother – Kouros assumed it was his mother – arrived to take away the protesting child. Kouros suddenly lost interest and turned to face his partner.

'I think we continue as we planned,' Farhad said. 'I'll continue to stay at the Johnson Hotel on Cerrillio and you stay here. We'll talk on our mobiles, and wait to see what happens. You think our professor will come through?'

'Sure of it,' answered Farhad.

'And what of the new American? The one with the white hat?' Asked Kouros.

'We wait. We plan. If he gets in the way, we both know what to do. Besides, he is old. What do we have to fear from him?'

There was nothing more to say. Farhad picked up his jacket, carelessly tossed on the bed an hour or so earlier, and without another word left the room, heading for the grey Pontiac and the close-by Howard Johnson. From the bedroom window, Kouros watched him go, smoking the last of his cigarette. It was an American brand, the last of his packet of Marlboro. He too turned from the window as the Pontiac drove away. Out of boredom he switched on the television. He pinched the formerly glowing end of his cigarette to ensure it was extinguished and then tossed it carelessly on to the otherwise spotless floor.

There was some mindless programme on the television Kouros believed he had seen as a subtitled version at least ten years before. It was called "The A Team". His mind was not on it. He was still thinking about "the plan".

Pahlavi had already told them about his forthcoming visit to Trinity Site. About it happening after his last day at the Los Alamos labs. Together, he and Farhad had looked Trinity up on the Internet. It was a long way from anywhere; the perfect place, part of the vast White Sands proving ground. They liked the irony of it.

In this most precious of American places they would kill Professor Pahlavi, and his protector too most likely. They would then take the two computer discs and the laptop computer. Maybe simply destroy them. Pahlavi had told them it was his last visit to the labs, that his calculations would be complete. They were essential to completing a larger plan the American Government were orchestrating called DEEP EARTH. The Americans would have no time to alter those plans. Without the disks what could they do?

If he had been allowed to proceed Pahlavi had told them he would simply have given the discs to his American contacts when he got back at Berkeley. They would know what to do with them. Pahlavi had added that the deadline for implementing DEEP EARTH, as he called it, was very 'tight'. Well, Farhad thought, it was going to get a whole lot tighter. It would be too late for the accursed Americans to do anything. It was why they had left their own actions until this very last moment.

The more Farhad thought about it all the more he liked it. And afterwards? He turned his own question over in his mind. If they got way with the killings at Trinity no one could ever pin it on foreign agency action or the Iranian government. If they didn't? They were willing to give their lives to the cause. They had no identity papers. They would die in an unmarked grave, happy in the knowledge they had wreaked the greatest of blows on their sworn enemy. It would be just recompense for what the Americans had done to their Bushr nuclear plant.

It was a good plan, for they need hardly do anything. And for the Americans, twelve million of them, they would suffer a mortal blow. Perfect.

Kouros decided he would go down the stairs, to reception, to buy more Marlboro cigarettes.

CHAPTER 15

For nearly two hours, Sir Freddie Stirling had been shown around GCHQ Bude, mostly in an army-issue Land Rover. Two other people had accompanied him: one a sandy-haired and bespectacled aide with an encyclopaedic knowledge of the establishment and the area's history; the other an army driver, who said virtually nothing.

During the above-ground tour, the civilian had included information about the village of Morwenstow's most famous inhabitant, the eccentric but Oxford-University-educated Robert Hawker. As the village's resident vicar for forty years, he had in 1834 begun the now widespread Anglican church custom of Harvest Festival, as well as composing what was widely regarded as the Cornish national anthem, 'Song of the Western Men'.

Before his arrival, Stirling had been expecting another Menwith Hill, the jointly operated GCHQ/NSA centre near Harrogate in Yorkshire, which he had recently visited. Covering six acres with its twenty-three spherical white radomes and three satellite dishes, it could track and decode five million message intercepts an hour, using its eleven underground floors of computer equipment.

Morwenstow was different both in terms of atmosphere and equipment. One difference had been the intensity of the artificial lighting. Paradoxically, Stirling had mostly noticed it for its unobtrusiveness. There were the rows of undistinguished Second World War-vintage single-storey brick buildings, and the airstrip on which three twin-rotor Chinook helicopters were standing while Loadmasters either loaded or unloaded them. As he watched, another Chinook came clanking in. There were four aircraft-sized hangars built in pairs, with one pair grouped some distance from the other. It was only when Stirling was taken inside one of the hangars that it became clear GCHQ Bude had a very different agenda from Menwith Hill. Hangar three contained not aircraft but rows of heavy-duty electrical switchgear, stretching away into the distance. Something, somewhere was drawing megawatts of power, Sir Freddie noted. He was told, too, that the power cabling linking it to the outside world had been laid underground to avoid unwanted questions from the curious.

Hangar two was dominated by a large steel-latticed tower that rose fifty feet above their heads, towards an obvious sliding partition in the roof. As he watched, two service lifts ascended or descended between the floor and a circular platform at its apex. Clustered around its base were a large number of monitors and recording instrumentation, reminiscent of a rocket launch facility.

'That's the transmission tower,' Sir Freddie's guide said, leaving him to muse over what that might mean, then added: 'They're checking the tower's optics.' He paused before continuing. 'Alignment has to be perfect, particularly for the full power tests.'

He waited for Stirling to take in what he could see around him before going on. 'A problem has been range limitations caused by the earth's curvature. One of the tests is aimed at eliminating that problem. We have a very simple solution.'

Sir Freddie was intrigued, and asked many questions as they walked back to the Land Rover parked at the hangar's door.

'How much power does the whole operation draw?' he asked.

'Twice the generating capacity of the entire country, but only for a short time,' came the reply. 'It's less than a trillionth of a second, but still a thousand petawatts.'

'A petawatt is a trillion watts?' queried Sir Freddie.

'Correct,' came the instant reply.

'It's the intensity that counts, I suppose?' Sir Freddie asked.

'Indeed,' responded his guide, surprised at Stirling's quick grip of the subject matter. 'I think they'll be ready for you downstairs now,' he added, checking his watch. 'We have our next low power test scheduled for 0432 hours. I understand you'll be attending, and they're anxious for you to know as much as possible before then. We'd better hurry.' They climbed back into the Land Rover and sped off.

Stirling had, so far, been impressed that such an establishment had attracted so little public attention over the years. Now, he was about to be shown its very heart, which he had deduced must lie deep underground.

Stirling's watch said 6.49 p.m. Over nine hours to the test, he said to himself, and the main reason he had been flown the two hundred miles from London. It was going to be a long night.

He turned the phrase 'the test' over in his mind. Everyone he had met at Morwenstow referred to the underground portion as 'the facility'. Even the intelligence briefings had referred to 'the facility'. So what was it, he thought, and how did it relate to what he'd already seen? Strangely, he'd never read anything about Morwenstow and, before he'd come, he'd looked for it on the Internet. There had been nothing; only a sparse amount relating to its GCHQ cover story.

The Land Rover's brakes squealed, bringing the vehicle to an abrupt stop outside a low-rise, nondescript brick building that could have seen wartime use. His guide jumped out, shouting over the noise and sharp coldness of the night air.

'This is our final stopping point for tonight, sir. I think you'll find what we have below ground is far more interesting than anything you've seen so far.'

Stirling quickly realised he had been fooled by the building's outside appearance. It was cleverly disguised to look old, but inside it was modern and hummed with low-frequency noise given off by something large and electrical, and out of sight. He could smell ozone caused by static in the air. It resembled a small airport departure lounge; its centre occupied by a bank of six or eight elevator shafts. Recessed into each panel housing, the 'call' button was a small video screen. He puzzled over them for a moment before one flickered into life. According to the screen, the lift was ascending from level 14: 'Cryogenics Storage Facility'. It arrived with surprising rapidity. Its brushed aluminium doors hissed open, allowing Sir Freddie and his guide into a capacious interior. It descended as rapidly as it had arrived, heading for level 17: 'Laser Optics and Superconductivity Device Assembly'.

Within a few seconds the doors opened again, to reveal a large and well-lit chamber with a number of corridors leading off, before each disappeared into the distance. Collectively, they gave the impression of a vast, underground complex, whose overall size he could only guess at. In the background there was a low-level hum, coupled with a vibration whose source was impossible to pinpoint. It didn't seem benign; it seemed darker, giving a sense of coldness ... of unease. Sir Freddie found himself shuddering involuntarily.

Over the next eight hours, with one stark exception, his time alternated between abject fascination and utter boredom. Leading experts in various branches of physics struggled to tell him what they were doing and why; the problems they had faced and overcome; their aims and their hopes for the forthcoming test. He understood it was all 'ground-breaking', and far ahead of anything 'even the Americans had'.

He was told about an electromagnetic pulse that always accompanied a nuclear blast, and had been observed initially with the first atomic bomb test at Trinity in Alamogordo; about the problems of 'death ray' particle-beam weapons, whether they were 'charged or 'uncharged'; about 'initial confinement fusion', and the technological strides made in pulsed power supplies of enormous output over nanosecond timescales; the role played by superconducting materials, capable of current densities in excess of 5 million amperes per square centimetre, but impossible to manufacture until recently.

Just as he was feeling weighed down by the intellectual and technical complexity, his guide whispered into his ear: 'Sir Freddie, I've been told that we have a surprise visitor to greet you.' Without waiting for a response he continued: 'It's Sir Robert Carr. He's expecting you. He's probably the world's most eminent mathematical physicist, and spent many years in the US, including Princeton and Berkeley. If it was not for him then all of this' – he gestured at everything surrounding them – 'would not be here.'

The guide continued whispering: 'You may remember that at the outbreak of World War Two, fearing the German development of an atomic bomb, Einstein was persuaded by fellow eminent scientists to write a letter to President Roosevelt. He warned the president about the possibility of the development of an atomic weapon of unimaginable power, which led to the formation of the Advisory Committee on Uranium, eventually superseded by the Manhattan Engineering District that built the bomb.' The guide paused, waiting for the information he had given to sink in, before adding: 'Sir Robert sent a similar letter to the penultimate British prime minister. The rest, as they say, is history.'

He began walking down the corridor, expecting Sir Freddie to follow. As he did so the ambient lighting switched from white to red.

'Don't worry, Sir Freddie,' the guide said over his shoulder, 'it's the night lighting taking over. We operate 24/7, and red lighting is easier on the eye. Takes less time to adjust to, and reminds the staff whether it's day or night back on the surface.'

Halfway along the long corridor were two black rubber swing doors. The guide halted by them. Sir Freddie peered through the glass windows inset into their upper halves and into the room that lay beyond. A lone figure was seated at a table furthest from the door. He was bent over slowly stirring what Sir Freddie imagined was a cup of tea or coffee but with an intensity suggesting he was oblivious to his surroundings.

Sir Freddie's guide whispered in his ear. 'If there was a Nobel Prize for mathematics then Sir Robert would have won it.' He paused and added, 'But I'm sure you know that there is no Nobel Prize for mathematics.' He pushed the door open for Sir Freddie to enter: 'He's expecting you. I'll come back for you in half an hour, if that's all right with you, sir.'

As Sir Freddie entered, the figure did not acknowledge his presence, did not look up until he was a few feet away. Then, the man suddenly smiled up at him. Overall, he had a kindly face, with bright piercing eyes. Sir Freddie was surprised to note that tears had been rolling down his cheeks. The man wiped them away with the back of his right hand, before withdrawing a large red-and-white-spotted handkerchief from his tweed jacket pocket.

'Sorry about that,' he said, with an almost inaudible BBC English accent. 'Bit of an emotional time for me. Not used to it. More the quiet of ivory towers of Cambridge ... that sort of thing,'

Sir Freddie sat down opposite and took in the man's face that, although heavily lined, was paradoxically young-looking, with twinkling bright eyes. Stirling thought he was probably in his late seventies.

'I suppose it is a bit emotional,' Sir Freddie agreed. 'All this is down to you, I've been told. With tonight's test it must be tense, wondering whether it will work.'

Sir Robert began to chuckle. 'First, I don't want to take credit where I don't think it's due. It was more that young chap Anderson's work. I merely checked his theory, the numbers, and said I thought it the work of genius. More importantly, I said it would work as forecast, given certain caveats. It was quite a short letter I sent the PM, really. As for tonight's test, I'm pretty certain it will perform as advertised. My current emotion is more to do with the law of unintended consequences.'

'The law of unintended consequences?' queried Sir Freddie.

In his quiet voice, Sir Robert responded: 'Yes. Those things we do with one set of objectives in mind, that end up spinning off into other unimaginable and often disastrous areas, for which we then, personally, have to pay. Maybe it's just the price of hubris, eh?'

He did not wait for an answer. 'Look at what happened to poor old Oppenheimer. The result of his work did the intended job, but what about the aftermath of contamination and deaths? We happily forget about those.'

There was a short silence. Sir Robert sipped his tea, leaving Sir Freddie unsure how to respond.

He tried to lighten the tone by saying, 'I don't think you need worry. We don't send people to the Tower any more.'

'You think not?' Carr shot a glance at him. 'Did you know Jonathan Anderson?'

' 'Fraid not,' replied Sir Freddie. 'I'm new to this, although I've heard of Dr Anderson, the role he's played, and then of course about his recent resignation.'

'Yes, his contribution was inestimable. His resignation was so unfortunate.'

'How long has he been gone from the Project?'

'About a month or two. He'd been troubled for some time before he went.'

'Oh?' Sir Freddie said, mildly curious.

'He's a religious man – not unusual with cosmologists and scientists of his calibre. They see things in the round, they see an awesome beauty in their work; a symmetry they believe can only come through the divine. His work began to trouble him. I could see it. I was surprised it took him so long to reach the conclusion he eventually came to. It had an echo of the old Manhattan Project about it, in that his work eventually came up against his religious beliefs. As the countdown begins, so these things tend to come to the fore.'

'It was the same for many who worked on the Manhattan Project, as the countdown began for the Trinity test and the inevitability of the atomic bomb's deployment against Japan. In retrospect you can more easily see their point: 40,000 people killed immediately in Hiroshima, with another 70,000 in Nagasaki. Altogether, including those that died later there have been to date 281,000 Hiroshima victims and 159,000 from Nagasaki. The Japanese call them the *hibakusha*. It literally means "explosion-affected people". It's an awesome cross we in the scientific community still have to bear.'

Sir Robert took another sip from his almost empty cup, then smiled at Sir Freddie. 'Conscience, of course, was not restricted to either the US or British scientists involved. Andrei Sakharov, father of the Russian hydrogen bomb, had it too. Maybe it goes with the territory, whenever scientists devise new means of mass extermination.'

Sir Freddie was beginning to feel uncomfortable. Sir Robert saw it.

'Please forgive my ramblings, Sir Freddie. Pity you won't be here to see one of our 'work-up' tests. You missed the last one we had on the twenty-second. The next one will be on the twenty-sixth, Boxing Day of all days. I think that one will be very late at night.' He paused for a few moments before continuing: 'On paper BRIGHTSTAR is like the old Reagan Star Wars or Strategic Defence Initiative. The difference is ... ours will work.'

'I think you mean you hope it will work,' Sir Freddie corrected.

'We know already. The individual components have been tested over the last two years. It's been tested at low power as a whole. In a few days we'll take it a stage further by trying it at a hundred per cent. But that will not be the only aim. Then it will be more to do with over-the-horizon targets, such as ship-launched missiles. Although not our principal target we have to be prepared to neutralise them, if I understand our latest government briefing correctly. We've solved the over-the-horizon problem by using a parabolic mirror aboard a specially equipped helicopter. The beam will be bounced off this mirror on to a target over the horizon. We're just testing the helicopter part of it tonight. It's more to do with command and control really. If it's successful, which I expect it to be, then the next test will find us using HMS Daring, a Royal Navy Type 45 destroyer, in a few days' time. She should already be on station somewhere in the Atlantic close to Fastnet Rock. More accurately she should currently be located in one of the MoD's gunnery and bombing ranges. She'll test-fire one of her missiles. BRIGHTSTAR will disable it despite it being "over the horizon", just as it would with any incoming nuclear warhead-equipped ballistic missile. It should mean that, within limits, the British will have a complete missile shield against incoming missiles, whether delivered from ship, land or air.'

'You think it's going to work?' Sir Freddie queried again.

'It will work,' Sir Robert replied, with a certainty Sir Freddie found reassuring.

'What made Anderson decide to jump ship then?' asked Sir Freddie. 'Does anyone know?'

'He'd become involved with an arcane area of mathematics, usually the preserve of cosmologists, and which he thought impacted his calculations. I later checked them myself. I could see his point, but thought his view was fanciful. Bit like is the bottle half full or half empty?'

'What did his latest calculations suggest?'

'In short, that BRIGHTSTAR might punch holes in the fabric of the universe. Theory suggests that this universe is just one of an infinite number. Every action that can occur, does occur: one where Hitler won the war, another where he didn't, for example. This universe is continuously branching to cover every eventual outcome. Everything that can happen, does happen in some parallel universe. Quantum theory, our most successful theory, never proved wrong, demands it!'

'So suppose he's right?'

Sir Robert laughed. 'He's not, and there's nothing to suggest he is. No evidence at all.'

'How would you know if he was?'

'Things would start disappearing and reappearing. We would need to have inadvertently created a gateway into multiple universes, each one slightly different from this. It could cause a problem. But this is the stuff of science fiction, Sir Freddie. We could be here forever discussing the "what ifs".'

'I take your point. But it all seems like science fiction.'

Sir Robert laughed. 'Science fiction, huh? One of Anderson's more fanciful beliefs was that the rocks, the earth surrounding us, were like a giant recording system, where the atoms of our bodies, and the atoms of everything else, are not so different when looked at from a quantum viewpoint. He believed that all actions are recorded by the local rocks surrounding that event; that somehow all actions impress themselves on everything else around them. So, for example, the floor and walls here carry a record of what we are saying now, and any other conversation that has ever taken place here. It was another area of concern for him that

BRIGHTSTAR might somehow decode these recordings. Very fanciful, very fanciful indeed,' repeated Sir Robert.

'My life is full of numbers, equations and symbols ... mostly symbols,' Carr said. 'For most they mean nothing, but to me they tell of collapsing black holes, of nuclear synthesis at the centre of stars, or even about this place. All I ask of the equations is that they are "beautiful".'

There was a noise of someone else entering the canteen. Sir Freddie half-turned to face the doorway, expecting to see his guide. Instead, he was surprised to see Sir Jack Geisner. Sir Jack walked over to join them and laid a hand on Sir Freddie's shoulder. He pulled one of the chairs out and sat down by his side.

'Sorry again, Freddie. But you know what our plans are. It's why you're here. Lots of last-minute snags to sort out. Bit of hand-holding. Nothing major. Usual run-of-the-mill stuff you get at a time like this, but they take time to sort out.'

'I understand completely, Jack. You've got to do what you've got to do.'

'Is there anything you want from me, Freddie?' Sir Jack asked. 'Anything I can do? You're in good hands here with Robert.'

'No. Nothing,' Sir Freddie replied.

'Good. Good. And Bob here is giving you all the information you want?'

'That and more, Jack.'

'This facility here is like a searchlight. With all that implies,' said Sir Jack. 'We have just the one projector. That's this one here. We don't need to build more. What's needed from here on in is a network of low-cost, simple receiving and switching stations, at key points around the country.'

As he was speaking, Sir Freddie got the impression he was being sold something.

'... they'll be nothing more than sophisticated parabolic mirrors to where we can transmit the beam from here, along a line-of-sight station, to wherever an incoming missile is expected. Automatic computer involvement means all the switching will be accomplished at astonishing speed. The result will be that any incoming missile, from any direction, can be shot down from the control room you saw on the surface. We've tested the theory out already, with an experimental station sited close to Glastonbury. Coincidentally enough, the beam runs along the same route as one of the old ley lines. Ever hear of those?' Sir Jack leaned back in his chair.

'Of course,' replied Sir Freddie. 'The so-called "lines of force". The stuff the Druids believed in; connected old centres of belief, churches ... that sort of old rubbish.'

Sir Robert chipped in. 'Might not be rubbish. Quantum theory links everything with everything. Ultimately, it links all times to one time. There is no past and no future. There is only this one time, and every possible time exists at this one time.'

'I think this is where I leave you again, Freddie, before it gets way above my head. I'm just a simple manager.' Sir Jack got up and began to walk away. 'I'll see you again shortly, Freddie.' He waved his arm and was gone.

'He's quite a character,' said Sir Robert with a smile, after Sir Jack's retreating form. 'He's the force one needs to get things done.' He stopped speaking for a moment before he added, 'I think I could do with another cup of tea.' He rose from his chair and added: 'Can I get you one as well?'

'No. I'm fine, thanks.'

Quickly, Sir Robert was back and continued where he'd left off. 'Now, where was I? Quantum theory.' He looked piercingly at Sir Freddie. Sir Freddie's grey eyes stared back.

Sir Robert seemed to have found his thoughts again. 'Some mathematicians think our own thoughts, involving billions of neurons, are arrived at through quantum processes. If they're right about how we think, then we're also linked to everything at all times. Makes us a bit like children of the universe. Humanity is inextricably woven into the very fabric of everything around us. It's what Dr Anderson was ultimately concerned with. It's something Freeman Dyson …'

'Who?' asked Sir Freddie.

'… one of our greatest scientists, but sadly living in the US for most of his life,' Sir Robert replied. 'He also seems to believe it, looking at some of the stuff he publishes with a religious bent.'

However interesting Carr's theories were, Sir Freddie, personally, had no way of assessing them. He trusted others had.

'I think if this "fabric" thing is true, it's more true for creative people. Artists, musicians, sculptors. Beauty is the mind of God. Their creative thought processes come out of quantum events.'

Sir Freddie could not answer. All he really knew was that he was in the presence of someone already in the history books as one of the 'greats' of British science. It was at that moment he heard the familiar voice of his guide from somewhere behind him.

'We have to go, Sir Freddie,' the voice said. 'There are a few more people to see. Sir Jack is most anxious you have the opportunity to see them all.'

'And I know what Sir Jack's like,' muttered Sir Freddie under his breath. He gave his thanks to Sir Robert and rose to leave.

As he was leaving, Sir Freddie paused a moment to look back. Sir Robert looked just as he did when he first entered the room, still hunched over, still contemplating his coffee cup. It was as if nothing had happened, that the time they had been together had never occurred. The only difference, if there was any, was that Sir Robert seemed somehow smaller and frailer. And Sir Freddie was left to wonder why 'the great man' had been crying when they first met?

It was something he could not dwell on. For shortly the test would begin, and that was why he had made the journey in the first place.

CHAPTER 16

The day at Pendragon Quay had been unusually warm, especially for December 28th. Christmas and Boxing Day had been good for the Anderson family, and this was their first outing since then. When they arrived with Ben's helper Dolores, it was 4.00 p.m. and the daylight was beginning to fade. A Super Lynx helicopter passed overhead. As they all got out of the car they watched it disappear towards the distant cluster of white radomes that was GCHQ Bude. Quickly the sound of its engines faded, leaving only the natural noise of seagulls, cormorants and auks wheeling on the wind, their cries mixing mournfully with the sound of waves crashing against the granite quay wall and outlying rocks. It was close to a high tide soon to be on the turn and known as 'slack high tide'.

'Didn't I tell you?' said Jonathan as he and Katrina waited for Dolores. 'Didn't I tell you it would be warm for the time of year?'

'Yeh,' replied Katrina. 'And it's about the 93rd time!'

As everyone got out of the car she went round to access Ben's wheelchair. In a few moments the four were ready for a short walk along the quay.

'Don't you think it strange we both had the same dream last night?' queried Katrina as they slowly moved off. 'I've been thinking about it since this morning. The chance of both of us having exactly the same dream is so remote. Then we both woke up with bad heads.' She paused a few moments, thinking before asking, 'How's yours now?'

'It's cleared. Like you, I took a couple of Paracetamol, and took two more just before we left to come here.'

'I almost feel fine,' Katrina countered. 'But not quite right; as if I've taken a drug. Like the aftermath of a sleeping pill. You know how they sometimes linger afterwards.'

Jonathan took hold of her hand gently as they walked across the quay and towards *The Smuggler's Inn* a hundred yards away. 'I don't know what it is, but I'm still not feeling quite normal,' she added. 'It's strange.'

They walked a few more steps before Jonathan said: 'A short walk in the sea air before we meet my old friend from university days will do us all good.'

'Richard? Richard Montanaro? The Reverend Richard Montanaro?' queried Katrina.

'Yes, that's right. I told you, we knew each other in our Cambridge Uni days. We had the same tutor in John Gill. Richard went off to work at the European Centre for Nuclear Research in Geneva, or CERN, if you prefer. He had strong religious beliefs, a bit like me. Later, he went off to work at the Large Hadron Collider facility. You must have heard of it?'

'Sure. Who hasn't?' Katrina replied. 'But let's finish the conversation about last night. Another thing is Ben. He was so quiet this morning. I've got this funny feeling. Don't know what it is; a bit like the old dark house routine where something odd affects everyone. Do you think I'm being silly?'

'You? Silly? You're not the type,' said Jonathan. 'Too rational, too calm. Besides, I've felt it too. Obviously, not as strongly as you have. Did Dolores mention anything?'

'No, nothing. She was sleeping in the lower part of the lighthouse while we were upstairs, enjoying the view across the bay. Wonder if that had anything to do with it?'

They had walked halfway to *The Smugglers Inn* when they heard the sound of an approaching car. The exhaust was loud, suggesting it was in need of repair. They looked back and up along the cliff road to see a car with a faint haze of smoke trailing behind it, making its way slowly down the zigzagging road towards them. As it drew closer they could see it was old and small and battered, its red body dented and scarred with many dark brown rust patches. Finally, it parked next to their own car. With a sigh like that of a dying animal, its engine stopped. A short, young man of around thirty-five got out, wearing a jacket with elbow patches, a dog collar and black tunic.

Jonathan and Katrina had already turned to start walking back in his direction. 'You'll love him, Kat,' Jonathan said in a low voice. 'We were great chums. And of all the luck, he ends up being minister here. At Morwenstow! Me being with the Project gave me the opportunity to come down a few times and catch up with him. It was an accidental meeting after I'd gone to the local church, what with it being so close to GCHQ Bude. That's where Richard popped up.' He shook his head. 'God works in strange ways. We'd lost contact a few years ago after one of our house moves; I'd lost his address. I last talked to him when we decided to come down here, and he's asked me to take the sermon at tomorrow's morning service. Hope you don't mind me working on our holiday.'

'Course not,' Katrina replied instantly. 'I like it when you do things like that. Reminds me of the man I married.'

'Hi, Richard,' said Jonathan, extending his hand to greet his friend.

Katrina instantly liked him.

'I've heard such a lot about you,' the Reverend Montanaro immediately said to Katrina, then changed the subject. 'I see you've brought the Army with you.' He laughed as he said it.

'What?' asked Jonathan.

'Over there, beyond the buildings, there's three Army trucks. They've set up some lights. I could see their personnel down on the beach. They look as if they're investigating the site dug by the archaeology team from Exeter over the past few months.'

'We didn't notice anything,' said Katrina. 'We were too busy looking at the sea.' She turned to look in the direction he'd indicated, but the activity was mostly out of sight from where they were standing.

'It looks fun,' said Katrina. 'Let's investigate.' Without waiting she set off in the direction of the one truck she could see. Before the rest could follow Reverend Montanaro shoved a book into Jonathan's hands. 'In case I forget,' he said, 'seeing as this is why we agreed to meet here in the first place, so I could give it to you. Be a shit if I forgot. Typical of me though.' He smiled. 'Forgetful as sin these days.'

He'd handed Jonathan a battered copy of Leon Lederman's *The God Particle*. Written in the early 1990s by former head of Chicago's Fermilab, a US national particle accelerator laboratory. The book had been proved prescient by later events. Jonathan glanced at it and muttered a mumbled 'thanks' as he inserted it into a leather 'man bag' draped over his right shoulder. They all then began following Katrina's receding form, with Dolores leading the way pushing Ben's wheelchair.

'Is she always like this?' asked Richard in a light-hearted manner as they set off. 'I take it you're gonna use the bit where Lederman links the Higgs field and boson to Genesis's Tower of Babel passage?'

'It had kinda crossed my mind,' Jonathan replied as they walked.

'And I've got to ask you what you found out with your dowsing experiment. When you first mentioned it I was fascinated by the possibilities.'

'Remind me about it when we're in the pub. Katrina is dying to know too, and we've not had time to talk about it.'

Looking over the wall they looked up the beach and could see the soldiers Richard had referred to. Within moments someone appeared out of one of the trucks and approached them. The white name tab displayed above the left breast pocket of his combat jacket identified him as a 'Captain Mark Lundis'. What it didn't say was that he was from British Army Intelligence.

'Be careful,' Lundis advised them. 'We don't know what we've found down there on the beach. It could be an unexploded sea mine. If it goes off it'll give us all more than a headache. Might even blow our hats off.' He was being good-humoured, but Katrina failed to notice.

'No, it's not a sea mine,' she said, with a note of conviction in her voice. 'It's not a mine,' she repeated with even more certainty; that she'd just made a statement of proven fact.

'We don't know what it is,' the captain exclaimed. 'Could be part of some wrecked ship's old boiler, or maybe a bomb. We have to be on the safe side. I'm sure you understand.'

'Have it your own way,' said Katrina, 'but it's part of an aircraft. The escape pod from something called AURORA. It says so along its side. "AURORA VI. High Altitude Escape Module", along with a flag of the United States. There were two hard-drive external computer discs on board. We thought they were amulets worn around the neck. They were inscribed "Property of the US Government". There's a hat too; a cowboy hat, a white Stetson, along with a bag of clothes, boots, and a check shirt. We took the hat to *Aquae Sulis*, the place you know as the Roman hot bath in Bath, we took them there so you'd know across fifteen hundred years...since the fifth century ... you would still know. There's a message... '

'I know you know it. You have to know it. Time is short. C'mon gang,' she concluded to everybody but still in a strange voice. 'No point in talking any further, if all he's going to tell us is lies. Let's go to the pub as we planned.'

Five minutes later they were sitting around a table in the almost empty *Smugglers Inn*. Dolores was helping Ben drink a bottle of orange juice. He was still quiet but his limbs seemed more under control than usual rather than having an independent life of their own. On the table was a round of drinks Jonathan had purchased for them all.

'Well,' said the Reverend Montanaro, breaking the short silence, 'what was all that about?' As he raised his pint glass to his mouth.

'I really don't know,' said Katrina apologetically. 'Something came over me, but I know I'm right. Don't ask me how I know. I just do.' Katrina looked at him. 'I'm sorry if I embarrassed you. I'm not like that. Really, I'm not.' She pulled an apologetic face.

'Here's the kicker though, Kat,' chipped in Jonathan as he finished taking a sip from his beer mug. 'I knew you were right. How could I know that? How could I be that sure too?' He looked at her, puzzled. 'We both had that same dream, of an ancient battle. I still remember it. So real; both of us agreed on that. You said over breakfast you thought it was of the Arthurian period. End of the fifth century?'

'Yes,' said Katrina. 'I remember, though it's grown fuzzy since this morning. You know, like dreams do.'

'Dreams, dreams,' said Reverend Montanaro, trying to be extra cheerful. 'Let's not read more into them than there probably is. Instead, let's drink to Jonathan and ourselves.'

Katrina and Jonathan looked relieved he had taken the recent incident with the army officer so matter-of-factly. 'I'll say this though, people,' Richard continued, 'what we've just experienced might well be a manifestation of the Higgs field and its associated boson. That's the thing me and a thousand others were going to spend our entire lives looking for when I was at CERN. The *Organisation Européenne pour la Recherche Nucléaire*. Ah! Those were the days.'

He saw a query form on Katrina's face. 'The Large Hadron Collider. The largest nuclear accelerator in the world, set up mostly to look for the mysterious Higgs.'

Katrina and Jonathan looked at him with renewed interest. 'Yes, people, if I may continue, one way of looking at the Higgs is it's the mechanism, the field that holds everything together; that stops solid objects flowing into one another, like me flowing into you.' He smiled. 'It's a supposed universal field stretching into infinity. It gives matter its mass. That's the short description. Some of us at CERN thought it was the universal field that could incorporate all other fields, as well as the history of all objects. So much so that inanimate objects can act like photographic paper or video tape, recording everything that happens to them and around them. Maybe you two have somehow picked up the vibes of this whole area. After all, it's said to have strong links with good old King Arthur and his Knights of the Round Table.'

Reverend Montanaro took another sip from his pint glass before continuing. 'It's all to do with the conservation of information. It's the same as the conservation of energy. Just like energy, one can never destroy information, only change it. Because of it, some scientists have taken issue with Stephen Hawking and his famed Hawking Radiation.' He stopped talking for a few seconds.

He could see that Katrina was staring at him. 'Go on,' she said. 'Is there more?'

'I don't want to get too technical. What I've just said was part of my old life. Like Jonathan's. Both of us thought science was not the complete answer to everything, as some think. There was still room for religion, still room for God. I know Jonathan and I share the same view as Freeman Dyson ...'

'Freeman who?' asked Katrina.

'Very famous British scientist who's lived in the States for fifty years,' responded Reverend Montanaro, 'famous for his work on the first atomic bomb, and later on quantum theory. But in that context, he said religion is bound up in the universe, and not separate from it. To my mind, it means that God is bound into the fabric of the universe too. Just as we human beings, who originated as star dust, out of the nuclear furnaces of stars, are part of the universe too. It's where God is also; actually bound up in everything we see around us. Not separate from it, like a man sitting on a cloud in the sky. But I digress from what Hawking said, and his conflict with the Law of Conservation of Information.'

'Where was I? Ah yes. The theory says you can't destroy information, but Hawking Radiation insists it must be destroyed. Others maintain it's held forever at a black hole's event horizon. They even maintain our whole universe is the inside of a black hole where all the information, everything that ever happens in it, you and me, every action ever taken, is stored as a hologram. That's a big, big thought, people, too big for me.' He sipped his beer before concluding. 'But it's fascinating stuff.' He paused, then added, 'And I've been talking too much.'

Reverend Montanaro put his finished pint glass down on the table. 'That's as far as I can go. Apart from the fact that if you accept the conservation of energy, the still unproven Higgs boson and its associated Higgs field, then stir in the concept of block time and you can see that we never escape from our past. It's with us the whole time. There is no past or present time … or even future time. It's all one.' He drew a deep breath. 'Now, what about another pint? Let a poor, hard-up Church of England parson do the honours.' He stood up and asked everyone what they wanted.

∞

Outside *The Smugglers Inn*, Captain Mark Lundis had been debating with himself about what Katrina had said. How could she know? he asked himself. How could she possibly know? Finally, he pulled out his mobile, punched in a number, and stopped walking down the sloping causeway leading to the beach and archaeological site.

A voice answered, and Captain Lundis explained why he was calling. There were long silences on the other end of the conversation.

'I don't know how she knows,' Lundis kept repeating. 'How the bloody hell would I? All I know is that they knew. Or at least she did.'

There was a pause.

'She did; the woman that's sometimes on the telly. History programmes, that kind of thing.'

Another pause.

'I don't have the authority to arrest them or anybody. And for what?' he said into the receiver, to his anonymous listener.

There was a short silence as he listened some more, before he added, 'How could she know what was inside the craft?' There was another pause as he paid attention.

'Since the students found it no one's been near. And certainly no one has opened it.'

There was another silence, as he listened again. 'Yes, I know what I'm saying is impossible. The hat, the clothes ... how could she know? But there it is. However, there were no signs of any discs she was talking about.' He pressed the 'off' button.

Captain Lundis was perplexed by what he now knew. It was impossible. It didn't make any sense. While he collected his thoughts, he watched the few soldiers still engaged on the Exeter site, collecting dead and scorched seabirds that were floating in on the tide and throwing them into the back of a small truck. There was already a small pile of forlorn-looking corpses.

∞

Reverend Montanaro placed the last of the drinks he'd just purchased down on the table, including a large glass of Californian white Zinfandel for Katrina. 'I'll have to save for a month on my parson's lowly pay to cover this lot,' he said as he sat down by the side of them. 'Now tell me, Jonny, just why did you come to Pendragon Point? And don't tell me it was simply for a holiday. That I just won't believe.'

But it was Katrina who answered his question. 'It's true, he did need a holiday after all he's been through at the Project, the bashing he took when he left. But I don't think we're breaking any confidences if I tell you he was after trying out a theory too.'

'Now we're getting somewhere. A theory you say. This is where I really come in. Sorting out theories is what I did at CERN as part of my job. People like Jonny here had the theories and we tested 'em. As I mentioned before, the biggest theory we were testing, and why the Large Hadron Collider was built, was to look for the "God particle", the Higgs boson. In experimental particle physics, they don't come much bigger.'

'I don't think Jonathan'll mind me saying that I don't think his theory was quite in that category, but it's still pretty big.'

'Keep going,' said Reverend Montanaro. 'I can feel the old juices rising!'

'We left with a couple of old wire coat hangers and that was it,' said Katrina. 'He was going to do a spot of dowsing. You know, wandering about …'

'I know what it is,' said a clearly excited Reverend. 'Don't tell me, Jonny. You were looking for a reaction from your dowsing rods. What you wanted to know was whether there was a link between those mystical forces of old, and the new forces you were invoking over the bay.'

'You don't believe in that old rubbish, do you?' Katrina asked aghast.

'Dowsing has been around for thousands of years. Bit like the Christian faith really. And both deal in things science can't measure. Yet a few of us old die-hards still believe, like we believe in miracles, and the Holy Spirit, and the power of God. We don't even ask for science to explain it. Actually, if it did, then many of us would think we'd lost rather than gained something. Faith is actually all about believing without proof. Proof is not required. We know it exists. That's enough. It's actually why I'm here, doing what I do. I say that … and I used to be an experimental physicist.'

'You're quite right, Richard,' Jonathan said. 'You and I share the same point of view. According to my latest theory, there would be a meeting of those forces of history and those forces out of Morwenstow. The theory, what I call my extension, said the very rocks of Pendragon Point would show traces of the forces Morwenstow were casting over them. My dowsing test earlier this morning was a way of seeing if that might be true. And yes, I know there might be other explanations. But in the time I've had, and the equipment I have available, it was the best I could do.'

'And what happened?' asked Reverend Montanaro.

'There was a reaction. The rods almost flew out of my hands.'

'So what do we conclude?'

'The Project is ahead of schedule and there's a chance I could be right.'

'What does that mean?'

'Not sure I can tell you, given the Secrecy Act.'

Jonathan thought for a few moments, before adding, 'There's a chance the Project will rip a hole in the space-time continuum.'

'That's bad,' said Reverend Montanaro. 'That's potentially very bad. I take it the blockheads in charge don't want to know.'

'That's it in one,' replied Jonathan.

'People, I've got to love you and leave you. A country parson's work is never done; people to see, places to go. And got to prepare for tomorrow's service.' He rose from the table. 'I know I don't have to say this Jonny, but don't let the buggers weaken you. And just remember, I'm coming round to pick you up tomorrow morning.'

'I've got to drive over this way anyway. Pick up a couple of infirm parishioners who don't have a car of their own. They live on a farm quite close to Pendragon Light. Are you joining us too, Katrina? Bit of religion is good for the soul.'

'Sure,' she said, smiling. 'We could leave Ben with Dolores.'

'Bring him too! I'm sure we can manage him,' Reverend Montanaro uttered. 'And afterwards, be my guest over at our local, *The Bush Inn*. Trust me, the food is to die for.'

CHAPTER 17

The Commanding General of the US Corps of Engineers looked uncomfortable. The president and Chief of Staff Cliff Wurzburg were sitting facing him over the table in the Roosevelt Room and could understand why. It had only been within the last few months he had been promoted from Lieutenant General and Chief Engineer in the US Army to his current position. Both the president and his Chief of Staff were aware that 'Chief Engineer' was largely a Pentagon desk job from where it was unlikely he would ever have visited the White House. But by all reports he was good at his new job. Chairman of the Joint Chiefs, Admiral Nelson W. Loadhammer, had said he was showing 'early enterprise and initiative'.

'Let me get straight down to business, General,' said the President. 'In a typical year your Corps of Engineers may have to respond to maybe thirty Presidential Orders connected with disaster declarations somewhere in the United States. These would normally be issued in connection with your supporting our Department of Homeland Security or the Federal Emergency Management Agency. Could even be both, depending on the scale of the disaster.' The president stopped speaking, allowing him the time to think further about his next sentence.

'Since New Orleans and Katrina,' he continued, 'we run emergency exercises all the time. We're gonna have one coming up in Washington State early in the new year. You've probably already heard about it. This meeting is to give you the heads-up that it's likely to be the biggest yet. So you'll have to pull out all the stops. We need you to work up a plan assuming a major earthquake in the area and impacting our former plutonium production site at Hanford. Anything affecting that site is gonna be messy. You've gotta contain it.'

The president stopped speaking, looking for a sign of reaction. His visitor's face remained impassive. The president continued.

'Obviously you'll be working with other federal agencies, which will probably include the Department of Defence. They're the only kids on the block with enough manpower to contain the disaster we envisage. The Chairman of the Joint Chiefs is already in on the loop.'

'We need a plan from you to us by February 1st with a "ready to go" date of March 1st. I know time is short. But emergencies often don't give us time to do all the planning we'd like.' The president looked at his watch. It gives you thirteen days from today to prepare your plan. One month to implement it. My Chief of Staff here will give you all the further information you may need. And before I go, welcome to the White House! I think it's a first time for you.' The president rose from his chair. 'I'll leave you with Cliff, here.' With that he was already walking towards the exit.

∞

An anonymous Gulfstream IV executive jet was landing on Kirtland's vast east-west runway. The pilot had chosen it in preference to the north-south runway because it meant he could taxi to its far end and be well out of sight of any prying journalists or photographers. Nothing on the external markings of the aircraft showed it was on long-term charter to the US Department of Energy.

Hoop Toberman was first down the aircraft's steps, followed closely by his usual retinue of staff, including a recently seconded member of the Secret Service carrying Toberman's 'football' chained to his right wrist. Few people knew why it was called a football. It was a briefcase carrying a laptop computer whose sole function was to act as a code machine; a replica of the one the president always had in close proximity and carried by an aide. For the president, it carried the codes necessary to launch a retaliatory nuclear attack on any country that might have launched a 'surprise nuclear first-strike'. In the past, this had always been assumed to be the USSR. The ending of the Cold War had ended that assumption. Now it was more likely to be a Middle Eastern country, Pentagon analysts assumed.

In Toberman's case, his 'football' held the codes necessary to break the encryptions guarding DEEP EARTH. Part of those encryptions covered the Los Alamos work carried out by Pahlavi. Like the president, wherever Toberman travelled his newly recruited Secret Service agent travelled too, carrying the DoE's 'football'.

Two anonymous black cars collected the 'energy entourage' almost immediately, only to drive them a relatively short distance to another part of the Kirtland base where a five-seater Bell JetRanger III helicopter was waiting with its rotors spinning ready for a quick take-off. Within a few moments all of them were on board with the helicopter climbing rapidly into the blue sky before heading south towards the Waste Isolation Pilot Plant close to the New Mexico town of Carlsbad 200 miles away.

WIPP had begun life in 1973 but since Congressional permission for its construction had been given in 1979, continuing delaying tactics by state and private environmentalists had meant it was still not as fully operational as originally planned. While everyone understood the problem it was meant to tackle, no one actually wanted such a potentially toxic scheme in their back garden, particularly when it might be there for the next 250,000 years. The same attitudes had seen the Yucca Mountain repository in Nevada also become becalmed.

But Toberman's mind was not on WIPP but on DEEP EARTH. He checked his watch nervously. Time was short and there was much to do before DEEP EARTH could become an actuality. He was under pressure to deliver. Luckily he had been divorced for twelve years and therefore had no excuses to make to any wife about 'working' on a Christmas Day or on a New Year's Eve in a few days' time. He was going to be busy.

He was well up to date on the key work still to be done with DEEP EARTH. But most of it was 'grunt work', he told himself. The question was where exactly to dig the tunnels and caverns, and where to place the nuclear charge? If they were in the wrong place then it could make matters a million times worse than not digging them at all. The equipment and men were already on site. All they needed was the word 'go' from him.

Pahlavi's Berkeley team had told him the final calculations would be ready 'within days', ready to feed into the Los Alamos Roadrunner simulation facility. So far, everything else was checking out. Radioactive waste shipments were already moving in and out of intermediate storage areas, with the shipping patterns prearranged so that any outside observer or whistle-blower would have little idea there was any deeper significance. They would see what they were meant to see and draw the wrong conclusions … as they were meant to.

So far there had been only two hitches and both concerned WIPP and The Environmental Protection Agency, under which a congressional mandate had said all limited nuclear waste shipments must be supervised. The result was the EPA was threatening to halt all such shipments almost as soon as they had begun.

Toberman had realised the real reason was small, and resulted from an initial minor misidentification of three waste cargos sent to WIPP from the now-closed Rocky Flats plutonium-processing plant outside Denver, Colorado. In itself it was not a disaster. But a relatively simple problem had been compounded further when a shipment from the Idaho National Engineering and Environmental Laboratory showed several key technical requirements were, according to the EPA, 'unsatisfactorily implemented'. Even after informing the DoE about the problem, several more shipments with the same problem had come through. That was why the EPA was threatening a national shutdown. The whole thing was probably down to a clerical error and had escalated out of all proportion.

Toberman reckoned that a show of due concern from the highest levels would quickly rectify the issues, and was why he was going to WIPP to meet EPA and state officials. He had promised the president as much. It was a louse-up for sure. It was simply Murphy's Law: if anything could go wrong then it surely would.

If that had been all there was to worry about then he would have ranked it as part of his 'everyday' concerns. But there had been another on the other side of the Atlantic. This one stemmed from a British company with long experience of nuclear clean-up and disposal, called Nuclear Waste Management. They were deeply involved in the US clean-up operations at many of the DoE's 128 ageing nuclear sites. But the British company had fallen foul of its own National Nuclear Inspectorate when alleged falsified papers had been discovered relating to leaks into local groundwater supplies from a radioactive storage tank at a plant in Cumbria.

Nuclear Waste Management's own estimate was that it had probably come from the previous operating facilities before their involvement with the clean-up had begun. But the issue had grown until it too was being threatened with total shutdown unless the British National Nuclear Inspectorate received a good and convincing explanation. But what did they mean by total shutdown? Was it likely to affect work the company was doing in the United States? And the explanations he'd received from the British made no sense however many times he'd asked. There was only one course left and that was a personal discussion in the hope of finding the truth. But he didn't want to merely kick the can a hundred yards down the road. He wanted to fix the problem, if there was one, for good. And he had to be at the next presidential DEEP EARTH meeting scheduled for early the following day. So the problem was how to be in two places at once?

As Toberman sat in one of the helicopter's rear seats he told himself that these things happened in any business. They were common in all systems where human beings were involved and, at some time, those human beings loused up. It was unfortunate that, in the British instance, those mistakes looked as if they could threaten the clean-up efforts in his own backyard, even DEEP EARTH, by halting shipments out of some of his own nuclear sites to Hanford. He'd had no option but to arrange a meeting with his UK counterpart. But there was only one way he could be there and attend the mandatory White House meeting. There was no way he could miss it, for he'd been briefed by Wurzburg that he was expected to reassure an increasingly nervous President that DEEP EARTH, under his management, was still on course.

The only way out was to use the Air Force's top secret AURORA. He'd used it once before so knew it could take him to RAF Fairford in Gloucester in under three hours. From Fairford it was then a quick helicopter flight to Cumbria on the UK's northeast coast, where everyone with anything to say was going to be represented. It was neat, too, since he could deny he'd ever been in the UK because of the timing issue. Even neater, AURORA's flight centre was right here at Kirtland air force base, a short hop from Carlsbad, where he'd be in

New Mexico. It was as if God had ordained it, Toberman thought.

Toberman went through the timing arithmetic again just to be sure. Take off time would be 1900 hours. Then allow for the six hour time difference between the US and the UK. Add in his estimated three-hour flight time. It meant he could arrive in the UK at approximately 0400 hours local time the following morning if everything went according to plan. His helicopter flight to Cumbria would take maybe an hour? The meeting itself? Maybe two to three hours? Flight back to Gloucester in one hour? Then the flight back to the US a further three hours with the provision of landing at Andrews air force base just outside Washington DC, and so cutting 2000 miles from his journey if he'd continued on to Albuquerque. And given that the time difference on the way back would be working in his favour? Yeah, Toberman said to himself, he could easily make his meeting on the 29th with the president.

Fairford, like Kirtland, had such a huge landing apron that no one would see them either approach or land, especially at this time of year, especially with its seasonably bad weather and low visibility. Toberman knew too that a C130J transport aircraft had lifted off several hours earlier with its service module, and was heading for the UK, ready for AURORA's planned arrival. Everything was set, he assured himself.

Three thousand feet below him he could see the reddish-coloured desert of New Mexico slipping by.

∞

It was seven days since Pahlavi had first arrived at Albuquerque, December 28th, and the fourth Saturday in the month. Santos was as good as his word. At 6.00 a.m., Robert Pahlavi's bedside telephone rang.

'Rise and shine,' said a familiar voice. When they met later in the lobby Santos was almost taciturn and perfunctory in manner, hungover from the previous night's heavy drinking. He apologised, saying he was never at his best early in the morning. It might take him an hour or so to 'warm up'.

The Jeep jerked away from the hotel and they headed south. US Interstate 25 mostly followed the winding Rio Grande river before it ended up in El Paso, Texas, 500 miles away on the US-Mexico border. Santos and Robert Pahlavi were not travelling that far. Journeying out of Los Alamos they retraced their drive of seven days ago, back to Albuquerque, on to the small town of Socorro, then Highway 380 at San Antonio, NM. It was shortly after this marker that Santos again noted the grey Pontiac. It was too far back to be certain, so he took the first opportunity to pull in, at *The Owl Bar and Café*, a low single-storey light brown blockhouse of a building. As he had expected, the Pontiac went on past but not before he had caught sight of its occupants, and positively identified them as Middle Eastern.

Later, over breakfast, Santos claimed *The Owl Bar* 'cooked the best green chilli burgers in New Mexico'. Munching his burger, he added 'San Antonio might look small, population just fifty, but it's where Conrad Hilton was born, 'an' he went on to build an entire hotel dynasty.'

Heading east along Highway 380, it was thirteen miles before they came to Stallion Gate, the only hole in a vast, chain-link fence. A sign announced it was here that the 'White Sands Missile Range' began.

The morning air was cold and bitter. They had seen very little traffic. Santos drawled: 'Don't let the traffic fool ya, son. With only two official open days a year, by 8.00 a.m. this road can look like an LA freeway at rush hour. But not today. Looks like it's gonna be quiet. Maybe nuclear bombs have lost their charm,' he drawled, 'huh?'

He drove up to the short Stallion Gate traffic queue, continuing his conversation. 'White Sands used to be one of the most godforsaken parts of the US. Ninety-seven ranchers used to own it before Uncle Sam moved in during the last war to use it as a bombing range. It was at a place called Alamogordo at 5.29 and 45 seconds a.m., July 16th 1945, they let off the big one with the world's first atomic bomb. It was equivalent to 18,600 tonnes of TNT.'

The traffic in front of them had cleared. Now it was their turn to pass through. Two armed guards motioned them to stop. They were businesslike in asking for the vehicle's registration certificate and the driver's valid driving licence to prove US citizenship. They also asked for proof of insurance. Santos produced his US marshall's badge and was almost immediately waved through, along with the cautionary words to drive carefully and be on the lookout for antelopes. 'They can be skittish and do unexpected things. Have a nice day.' One of the guards shielded his eyes from the sun as he watched Santos drive off up the road.

Heading south-south-east they came to a left turn at the 'Permanent High Explosive Testing Site', its small group of trailers clustered around a domed building. As they drove up, another armed guard, standing conspicuously in the middle of the road, indicated a left turning.

Heading east towards the Sierra Oscura mountains, they passed original artefacts from the Trinity test. Among them they noticed the original wooden posts still carrying the remains of long disused instrumentation cabling. Santos explained it had once carried monitoring signals from Ground Zero to recording devices housed in a still visible earth bunker they were just passing.

In the main, Santos thought Pahlavi had been more quiet than usual throughout the whole journey. He took it as an ominous sign, for he thought he would otherwise have been fascinated by what they were seeing, given his initial interest. On the other hand, it was difficult to know whether Pahlavi was simply overwhelmed by the endless stretches of blindingly white sand forming the largest deposit of pure gypsum on earth. Or maybe it was the enormity of what had happened here the best part of a hundred years before? Either way Santos would be maintaining his vigilance for the coming day. Whatever happened, he had to make Kirtland by 6.35pm that evening, at the latest, in order to babysit the Secretary of Energy.

A white car being slowly driven in the opposite direction passed them, bearing the legend 'White Sands Missile Range Security Police'. There were three men sitting inside.

Then Santos came to another checkpoint. They were stopped by more uniformed guards only to be handed several leaflets, one on the history of Trinity site. They passed another sign insisting: 'No Stopping. No Drinking and No Smoking within the inner fenced area'.

Pahlavi scanned the leaflets. They included two giving the warning 'No dirt or any other material is to be taken away', along with the more mysterious 'No applying cosmetics within the Ground Zero area'.

Sixteen miles further on they came to a soldier standing in the middle of the road. He waved them to pass on his left, past a derelict concrete bunker that one of the information leaflets Pahlavi had been studying explained: 'had once contained measuring instruments for monitoring ground radiation after the blast'. A few yards further on they came to a large, circular chain-link fence area with the single sign saying starkly:

Ground Zero

'Here ya are, Robert. Not much to see today,' said Santos cheerily as they both got out of the Jeep and began walking over the dry, white sand. A small part, closest to the track that had led them there, had been covered by the US Corps of

Engineers, with crushed rock and aggregate, to make a firm car-parking base.

Santos was deliberately trying to sound more casual than he really felt. A few moments ago he had rediscovered the now dust-covered grey Pontiac flatbed truck parked haphazardly in the car park. He'd learned not to believe in coincidence. Someone coming from Albuquerque airport and travelling along the same road to Los Alamos was an acceptable possibility. But to Trinity Site as well? It stretched 'coincidence' a tad further than he was prepared to accept.

But where were the Pontiac's occupants? They'd be dangerous and prepared to kill. The death of an NID agent in San Francisco was a pointer. There was nothing definite; nothing positively linking them to his death ... just a feeling. But if it was right then there had to be another possibility too. Robert Pahlavi had to be in on it. And all of them were trying to take him for a patsy. Not only that, they were taking the US government for a patsy too. He felt the hairs on the back of his neck tingle with anticipation. More prosaically he felt the lump of his Glock 22 tucked underneath his left armpit.

Act as if everything is normal, he told himself. You may be wrong. Act as if everything's normal, but be prepared for anything.

Santos said as casually as he could: 'There's mostly just endless, flat sands with that' – he pointed a finger – 'at its centre.'

A short distance beyond where they were standing was a twelve-foot-high, triangular black obelisk made out of volcanic stone, standing in a surprisingly shallow depression. It was the major, tangible evidence of the US Corps of Engineers, who had filled in the substantial hole caused by the atomic bomb's detonation in 1945, as it sat atop a hundred-foot metal tower.

For half a mile around where they were standing, the desert sand had been turned by the heat from the blast into a translucent green, glass-like substance, later christened 'Trinitite'. One of Pahlavi's pamphlets warned: 'Little of it remains … and its removal is forbidden'.

They walked over to read the obelisk's black inscription:

> *Trinity Site*
> *where the world's*
> *first nuclear device*
> *was exploded*
> *July 16th 1945*

A further sign underneath stated:

> *Trinity Site*
> *has been designated a National historic landmark.*
> *This site possesses National Significance in*
> *commemorating the history of the*
> *United States of America 1975.*
> *National Parks Service*
> *United States Department of the Interior*

Santos checked his watch. It was 11.30 a.m. Where were those two motherfuckers from the Pontiac?

They walked all over Ground Zero in little over two hours. They inspected the rounded, circular features of a steel mock-up of 'Fat Boy', the plutonium-based atomic bomb actually used for the Trinity Test. It measured a little over twelve feet long and eight feet in diameter, and was presented on a replica of the trailer used to carry it to the test site.

There were carefully signposted picnic areas with selected items of debris for the interested. One picnic area displayed a section of pipe thirty feet in diameter, whose steel walls were eight inches thick. A sign said that in the event of the bomb failing to explode as intended 'the steel Containment Vessel' was meant to prevent the bomb's plutonium core contaminating the landscape. Later it had been blown apart by large charges of TNT.

They continued walking around the desert, kicking the occasional stone, reading notices, studying faded photographs pinned to fencing. They all told similar stories about how safe Trinity was now. Aircraft passengers suffered a larger radiation dosage than any they would obtain by walking around Trinity. Geiger counters, placed strategically at the entrance to Ground Zero, could prove the point if anyone wanted to know.

Eventually, and in silence, they wandered back to the Jeep. For a few moments Professor Robert Pahlavi shielded his eyes from the light and slowly turned as he surveyed the whole of Trinity. It might have been a natural movement for someone overawed by Trinity's historical significance. Or it could have been that he was looking for someone.

Santos noted the Pontiac still parked 200 yards away, one of only three vehicles still left in the car park. And one of those 'other' vehicles was their own. It was easy for Santos to deduce to whom the third car belonged. A family of four with two young children were still playing in the sand of the 'play' area, ignoring the alternatives of expensive swings, see-saws and roundabouts.

Typical kids, Santos thought as his gaze passed on. But there was no immediate sign of anyone else. Then, out of the corner of his eye, he saw his quarry emerge from behind the family car and climb into the Pontiac. Where's the other guy? Santos thought.

The Middle Eastern-looking stranger drove the Pontiac slowly round in a large circle so that it could be driven back along the track eventually leading to Stallion Gate.

Santos glanced at Pahlavi. He looked sombre. His mood had changed. He was nervous too, glancing furtively around him, as if looking for someone else to be there. He was using a large handkerchief to mop his brow even though the air temperature had fallen substantially to be cold enough for snow.

'Are you okay, Prof?' Santos asked again as casually as he could.

'Oh, yes. I'm okay. Really. Feeling a bit nauseous, that's all. Must be something I ate.'

Santos let the remark go. The professor was no soldier. He was expecting something to happen, and suffering from nervous tension.

Years of doing the same and similar jobs told Santos to be on his guard, even if he hadn't been before. The trick was to still appear cool.

Santos picked up on a conversation they had been having earlier, hoping it would calm the Professor.

'Do you see my point?' Santos heard himself saying.

'No,' said Pahlavi flatly and, Santos thought, nervously.

'That's the point,' said Santos triumphantly. 'The whole area has been sanitised, swept clean. It's like Disneyland. It's public relations, proving how safe nuclear energy is. On these open days each visitor goes home to tell neighbours, relations and friends where they've been, and what they've done. They say, "We ate hamburgers where the first bomb went off, so how dangerous can nuclear waste be?" It's public relations, designed to allay John Doe's suspicions that someone's been shitting in their backyard for the past seventy to eighty years.'

'What they don't know about Trinity is that it took twenty years to clean it up so it looks the way it does now. They took away millions of tonnes of radioactive desert so they could present this Disneyland picture.'

They climbed back into the Jeep and as he fired the engine up and began to slowly drive back along the dirt track, away from Ground Zero, Pahlavi said in a tremulous voice:

'There's one place we haven't visited yet.'

'Oh yeah?' Santos replied, pretending not to know. 'Where's that?' He, of course, knew the answer to his own question. There could only be one place: the McDonald ranch, the place where Oppenheimer's team had first assembled the Bomb.

It was two miles back along the track that had led them to Ground Zero, far enough off the main track to avoid many visitors. So that's where they're gonna try and dry-gulch me, Santos told himself. Well, we'll see about that. We'll see who's the patsy. He thought very briefly about phoning for back-up. Did he need it? Naw. He could take a coupla kids anytime. And what about the professor? He could take his chances for selling Uncle Sam down the river.

Yeah, it was gonna be the McDonald ranch. He should have seen it coming earlier. That was where one of them had already been deployed. Now, the guy he'd just seen was joining his partner. It was a pretty clear plan, and simple too. McDonald was the natural place for a set-up to be staged. He would just have to take it as it came. He had no other choice. And he could, even now, always be wrong. But what 'they', whoever 'they' were, didn't know, was that he'd had been to Trinity a million times before. He knew every inch of it, particularly the McDonald ranch; had even arrested one or two bail-jumpers at Trinity in the past. He knew where the 'set-up' simply had to be. Santos looked at it as if it was going to be the 'Gunfight at the O.K. Corral' where the Earps had met the Clanton gang back in 1881. The McDonald ranch was going to be his chance of carving out a little bit of history.

If things got outta hand, he could always summon either the guards from Ground Zero, or those at Stallion Gate. There were the mobile units too, one of whom they had passed earlier that day. He could summon any or all over his sat phone. He'd warned White Sands police the day before, but asked them to stay well out of it, unless asked. He wasn't that old. He was still well up to any shoot-out.

Built in 1913, by German immigrant homesteader Franz Schmidt, the more accurately named Schmidt-McDonald ranch house was on a narrow dirt road. The Schmidts had sold the property to the McDonald family some time in the 1920s, after which they had gone to live in the more equable climate of Florida.

Farhad had been at the ranch for most of the day. He wanted 'some action'. Any action would do. He'd looked out of the window to the distant snow-capped Jemez mountain range more times than he cared to remember. For want of anything better he'd read and reread the information leaflets about Trinity and the McDonalds' house.

It was a single-storey mud-brick and corrugated-metal-roofed building about 2,000 square feet in size. An old icehouse and cistern had been built on the west side, with a stone extension added to the north wall containing a modern bathroom.

To the east of the house was a divided concrete water storage tank, windmill and deep well with a low stone, circular wall running round it. Further east were the corals and pens for livestock. From his position looking out of the window he could also see the remains of a bunkhouse and barn, where the Pontiac was currently hidden.

Kouros had joined him fifteen minutes earlier to tell him their quarry was on its way. Mobile phones were out of range of any cellular tower. Kouros would be ready, he told himself. They would both be ready. What had they got to lose?

He heard the sound of the Jeep first. The noise of its engine carried far in the still air of the desert. Strangely, Trinity reminded him of some parts of his native Iran.

CHAPTER 18

The president paced the deep blue carpet of the Oval Office. 'I don't know how long we're gonna be able to keep DEEP EARTH quiet,' he said to Cliff Wurzburg. 'For the back-up plan, I'm already having to get more and more people involved. You know what Washington's like? Leaky as a sieve. I'll be reading about it next in *The Post*.'

Wurzburg was sitting calmly on one of the room's sofas while the president, showing the strain he was beginning to feel, continued to pace backwards and forwards.

'I shouldn't worry overmuch, Mr President. These guys are used to keeping secrets,' responded Cliff Wurzburg simply. 'One more secret among a basketload of other secrets is nothing to them.'

'I hope you're right.' The president did some more pacing before adding: 'It's 28th December. We've got thirty-four days left. Till the end of January. Is it going to be enough?'

'Sure it is, Mr President. We have good men on the job. They're all experts at what they do,' replied Wurzburg. 'Everything is still on schedule.'

∞

The roads were deserted on the drive back from Alamogordo. Santos made good time, even though a light snow had been falling. Even into Albuquerque the roads remained quiet, apart from occasional gritting lorries. The Jeep eventually turned into the barriered main entrance, past the large, orange sign proclaiming 'Kirtland Air Force Base, Materials Air Command'. An armed sentry, one of two, came round to Santos's side of the Jeep, ready to inspect his papers.

Within a few brief moments he was driving through. No questions asked. But then he'd expected none, for he was security cleared to the highest level. The White Sands police were sorting out the mess he'd had to leave behind. One man dead; one man badly wounded, but he should live; and one man, the professor, with a flesh wound, a bullet graze, nothing more. He was gonna have plenty of time to recover in a state penitentiary on charges of state espionage. Meanwhile he would have to wait in the local Albuquerque prison cells until he got back from the UK and his protection duties involving Secretary of State for Energy Hoop Toberman were over. Aboard AURORA his last-minute trip to the UK should last less than 24 hours before he was back in the US he told himself.

On the passenger seat next to him was the professor's laptop, together with the money belt holding the two disc drives. He'd checked them: 'Property of the US Government', they said. So he'd make sure they were handed back. But at the moment there was no time. He would have to take them with him on the flight. But first he had to get to Kirtland, but quick. He checked his watch. It said 1835 hours. Take-off time was scheduled for 1900 hours. It was gonna be a close-run thing, maybe just minutes to spare before they were on their way to the UK by fast jet ... a very fast jet.

As he drove slowly on to the air base, in his mind he ran through the events of the day. It had all gone relatively smoothly. He'd first realised they were no longer being followed once they'd left the Owl Café that morning. That could only mean their tail had gone on ahead, which meant they had to know beforehand where he was heading. At White Sands Stallion Gate, and out of earshot of Pahlavi, he'd asked about any other visitors; whether there had been two men of Middle Eastern appearance. He'd explained he didn't have any pictures. The NID were going on eye-witness statements. To clinch it, he'd described the Pontiac. 'Yeah', they had said. 'It's been through.'

When they'd arrived at the flat, open desert space of Ground Zero there was no one there. Deduction said the 'operation' if there was going to be one would have to be at the McDonald ranch. There was nowhere else for it to be.

From there on in they'd walked into the McDonald ranchhouse to be met by two guys with guns. He'd used Pahlavi as a shield, reckoning that as he was in on it all, they wouldn't shoot. When they did, he'd thrown Pahlavi one way, while he'd dived the other, reaching simultaneously for his Glock. Their first shot had missed Pahlavi completely. Then they got off another. He'd been puzzled about their aiming at the professor and not at him. It was academic. He'd been lucky while their luck had run out. He'd got two shots off in quick succession and they'd both found their targets. As they'd fallen their guns had fallen from their hands to clatter on to the floor. Afterwards he'd only wished all his cases had ended as simply as this one had.

∞

Aboard *HMS Daring* it was 29th December, 2305 hours. They were almost four hours ahead of schedule, which was 0345 hours. Over the ship's intercom had come the voice of the ship's captain. It said they were nearing their rendezvous point and would be on station within three hours and twenty-two minutes. Everyone in the ops room listened intently, not wanting to break their concentration on the action data display screens in front of them. The old hands had done similar exercises many times before. Usually, they did everything they were expected to do in times of actual warfare, except a live missile firing. Government budgetary constraints meant it was a rare event since each missile cost over a million pounds. This time was different. Just one missile, they'd been told. But it was going to be the 'real thing'. It was enough to get everyone's adrenalin going.

Santos made it to the Kirtland air force base 'AURORA Facility' shortly before 1845 hours. There was the formality of going through another security check, a procedure that basically consisted of checking his name off against the one already held on the AURORA flight manifest. An air base car then took him the not-short distance to the aircraft waiting at the furthest point on the runway and hidden from view by vast hangars.

∞

AURORA's pilot and co-pilot were already seated in the aircraft's cabin where they had been for several minutes going though the few preliminary flight checks they still needed to carry out. The two-man crew wore a dark grey, all-in-one flight tunic, with their names emblazoned on a white square over their left breast pocket. The pilot checked his watch again. He was used to passengers showing up late, sometimes not at all. His watch said 1850 hours Mountain Standard Time.

The co-pilot was the younger of the two, a 37-year-old married man from Columbia, the state capital of South Carolina. Both had been handpicked to fly AURORA for it was like no other aircraft on earth. Originally, it had begun life as a photo-reconnaissance design at Lockheed Martin's famed Skunk Works in Burbank, California. It flew at Mach 6, at a height of 200,000 feet, using stealth technology, so was virtually invisible to radar. But the end of the Cold War meant it had no further objective in that role. $400m had already been spent from a joint 'black' USAF/CIA budget, just as it was going into production. To save face among certain senior air force staff its specification was changed to accommodate a total of six people, including aircrew, who could then be flown anywhere in the world at a moment's notice, and arrive within three hours, after allowing for in-flight refuelling. Although used mostly by the military and the CIA, AURORA was sometimes used too, as on this occasion, by politicians in a hurry.

AURORA's co-pilot had, for most of his service life, flown the supersonic Boeing-North American B-1B Lancer, a variable geometry Mach 1.5 bomber that, together with the stealth Spirit and the venerable B52H, were the backbone of Strategic Air Command. Although the eight-engined B52H had first flown in 1954 it was planned to remain in active service life until 2040.

All AURORA'S pilot cared about was that his co-pilot was as good as he was. And when the chips were down he would do his job. Some cost-cutting Pentagon based accountants had argued a co-pilot was unnecessary. But they didn't have to fly it. Emergencies happened. When they did, they could happen fast. That was when an extra pair of hands was essential; the difference between life and death. With AURORA, emergencies would come fast and high. In reality, there might be little anyone could do except fire the explosive bolts holding the long nose and passenger section to the rest of the fuselage. Then it would be down to praying the aircraft's designers had known what they were doing when it came to the escape module.

An additional safety measure was the light blue coloured flameproof flight suits and helmets along with integral compressed air supply the aircraft's designers had thoughtfully specified. Passengers were supposed to wear them, but most never did because they looked hot, uncomfortable and cumbersome. The usual exceptions were CIA operatives, who wore theirs as a matter of course. It was a part of their training, the flight crew supposed. Some came already wearing their tunics, others carried them on board, packed neatly in a small brown bag embossed with their individual name, and withdrawn from 'stores' shortly before each flight.

The biggest risk to AURORA was when she was losing altitude and speed, otherwise she was invulnerable to missile attack. But when she was decelerating it could be different, even fatal. All the crew could hope was that stealth technology and limited electronic countermeasures would render emergency manual procedures unnecessary. Besides, she would usually only be losing speed over friendly territory, unless there was an unfriendly 'Indian' out there armed with his own RPG and his own terrorist agenda. It was a possibility in certain Middle Eastern terrains. But the history looked good. The old SR-71 Blackbird had flown over hostile territory countless times during a 34-year period, and one had never been lost due to enemy action.

The flight crew made small talk while they went through the pre flight checks. The pilot looked again at his wristwatch. There was four minutes to go. The flight plan was already filed with the crew expecting the forthcoming flight to be as routine as countless others had been over the two years they had been cleared to fly AURORA.

For all her technical sophistication she was a docile aircraft to fly, provided strict attention was paid to flying her within her intended operating envelope. Otherwise, her inbuilt airframe instability could yield sudden and disastrous consequences. Flown correctly, she had no bad characteristics any experienced pilot needed to guard against. In some respects, she was easy because her original design specification had only called for her to fly high and fast.

The pilot was ex-US Navy with twenty years experience. That experience had taught him that there were old pilots and there were bold pilots but there were no old, bold pilots. Afterwards, he had been asked to join an air force special operations unit, testing experimental aircraft, such as the YF-22 before it became the Raptor, and based at Edwards Air Force Base, one-time home of legendary aircraft such as the

Northrop X-15 rocket plane, now hanging proudly in Washington DC's Smithsonian Museum. In the 1960s, this had broken more world records than any other aircraft, with its ability to fly in excess of Mach 7, at heights of 70 miles and a range of 300 miles. Its part in the ten-year X plane programme had contributed significantly to the later Space Shuttle, a craft that regularly achieved speeds of over 17,000 mph, an operational height of 600 nautical miles, carrying up to ten passengers in addition to its flight crew of six.

But a natural outcome of the space shuttle's 1970s-based technology was the support infrastructure and the sheer manpower needed for each flight preparation. The weather, too, had to be just right, or a mission would be cancelled at a moment's notice. If it was, then it could take days, even weeks, to prepare for another.

AURORA was different. Preparation and pre-flight checks were more akin to a conventional passenger aircraft. She even shared some of her construction technology with them, in the use of plastic composites now used in the Boeing 787 Dreamliner, and the EADS A380 Superjumbo.

All in all the pilot considered himself lucky to still have a job. What else could a 51-year-old ex-Navy test pilot do? Flying AURORA was not quite as far on the wild side as it had once been, but it still took a lot of beating. Whereas astronauts might only fly into space once or twice a year, he did it three or four times a month, sometimes more.

There was, of course, the return journey to be considered. To cater for that, the Air Force had developed a shrink-wrapped, air-transportable and containerised flight support package. It could be carried in the 4,500 cubic foot hold of any of the Air Force's 150 C-130J Hercules, to any one of a number of pre-selected strategic points around the globe. RAF Fairford in Gloucestershire, close to the town of Cheltenham, the home of GCHQ, was one of them. A C-130J had flown out to the UK the previous day, ready for AURORA'S arrival.

AURORA regularly flew to the fringes of space and could go beyond if necessary. In appearance she was a cross between the delta-winged, 'Orion' spacecraft seen in Stanley Kubrick's cult film *2001: A Space Odyssey*, and the now retired space shuttles. AURORA was smaller and, after fifteen years, was still classified and largely unknown to the outside world.

AURORA'S most secret aspect lay in the design of her engines. They were originally of British Horizontal Take Off and Landing (HOTOL) design, were air-breathing up to 90,000 feet, before switching over to rocket propulsion during the final leap into a suborbital flight path. There were five other sister craft of which two were constantly being serviced. The remaining three saw regular use after somebody, somewhere, had reasoned that, in the wars of tomorrow, senior military staff mobility, along with command, control and communications, would be the new frontier. Despite modern technology, it was still essential to have a Mk1 eyeball in the right place and at the right time, in any theatre of war. Or, as AURORA's critics on the Congressional Oversight Committee would say: one of the Pentagon's most expensive procurement programmes had ended up becoming nothing more than an executive jet, ferrying around top-level Brass. Even politicians.

According to the crew's flight plan, they were expecting three passengers: Energy Secretary, Hoop Toberman; his assistant; and the other was an unnamed latecomer whom the pilot assumed from past experience, would probably turn out to be CIA. They were always latecomers and told them nothing. He never asked. He'd been briefed not to. All he needed to know was whether he'd been cleared for flight by 'security'. Suddenly he heard over his headphones that the airport car was approaching with one of his passengers and a helicopter would be landing shortly, in less then five minutes, with his remaining two passengers.

The runway itself had already been cleared and it had stopped snowing. It was a good sign the pilot thought. He saw the car approach and stop a few metres from his aircraft. A man got out. It was a short walk for Santos before he was ascending the ladder leading into her. As he entered he glanced up at the small, white lettering stencilled over the entry hatch, above the United States flag:

AURORA VI
High Altitude Escape Module.
United States Air Force
CXFM130508

Santos had seen it all before. Inside, it was small and cramped, with only enough room for four seats apart from the two-man crew. The passengers were separated from the crew by a bulkhead, through which there was a small hatch. He knew that during the flight there would be little for the crew to do, and that the two-man crew hated their enforced inactivity. Santos had talked to them often on previous flights. For them he knew the real joy of flying was to be near the edge, when it was just them, the machine and the laws of physics. Neither minded jokes about being 'rocket jockeys'. Then there was the view or the earth outside; something without equal. From the fringes of space, the beauty of what lay, both below and beyond, never failed to be both humbling and amazing. They felt privileged to be reminded of the awe and splendour of the universe most had told him.

AURORA was ready. Where were the rest of his passengers the pilot asked himself? Outside the temperature had continued to drop. It was going to be a bitterly cold, starlit night. All unnecessary personnel had been withdrawn from the area, and the pilot could see the hydrogen fuel used in the earlier stages of the flight, slowly venting and turning into what looked like steam. He checked his watch again. It was 1858 hours with take off scheduled for 1900 hours. So where were his two remaining passengers?

The pilot then heard the sound of an approaching helicopter. A Bell 206B JetRanger III dropped suddenly into view. As it touched down, two men jumped out and ran the short distance from it towards the awaiting AURORA. Both carried briefcases, one of which was chained to one of the men's wrist. They scrambled aboard amidst apologies for being late and gasped explanations of 'the damned weather ... thought we were never gonna make it.'

The pilot waited for his passengers to remove their coats and settle into their seats before shouting back to them.

'Sorry, folks. For those of you who've not travelled with this airline before, the bad news is the seats are uncomfortable. The good news is you won't be sitting there long. Now buckle up and we'll git movin'. Next stop the UK, and we won't spare the horses.'

He waited a minute or two more before giving the order for the tow truck to move AURORA out onto the runway, clear of the hangar. He watched the tow truck disconnect and lumber away.

'Are we okay back there?' the pilot shouted through the hatch.

'Sure. Let's get the show on the road,' came back a voice.

The pilot checked his watch. 'Bang on the mark' he said to himself. Followed by '1900 hours'. Then more loudly over the roar of the engines. 'Okay folks. Here we go.'

He moved the throttle levers slowly backwards. The start truck that had been injecting fuel into the inlet manifolds moved rapidly away. The roar from the engines increased followed immediately by a tremendous vibration that shook the whole aircraft. AURORA began to move slowly to the end of the runway before turning. Flight clearance came almost immediately

Hoop Toberman looked up and saw a red light change to green.

'Stand by,' came the pilot's voice. Moments later came the okay from the tower's flight traffic controllers. The pilot lit up the afterburners.

'Lighting up, folks,' he shouted.

Flames shot several feet out of the engine exhausts. She was away. For two of the passengers it was like a sudden punch in the back. For the other he was dragged forward against his seatbelt.

Toberman stared across into the face of Tom Santos. The face was serene while his own, he guessed, must've looked petrified. He saw a hand reach across the gap separating them. 'Hi, Mr Secretary of Energy. The name's Santos, Tom Santos and I'm here to look after you until we arrive back in the US.' The Secretary limply shook the extended hand. Santos noted it was like shaking a long dead halibut as he simultaneously felt the butt of his own recently used Glock underneath his armpit.

At 234 knots the pilot yelled: 'VI,' followed shortly by 'V2 Rotate.' AURORA left the runway accelerating hard. Rapidly the runway fell away as she climbed faster than any conventional fighter. In ten minutes she would rendezvous with a KC-135 tanker at 35,000ft, for AURORA had taken off with the minimum fuel in order to keep her weight down and so give her the maximum of range.

In twenty minutes, she would be just below supersonic speed, flying east along the military air-lane passing north of Dallas. Then, over the states of Mississippi and Alabama, before flying out over the Gulf of New Mexico. She would then turn north to track over the Atlantic at Mach 6 or just a little short of 4000mph. This was not the speed she flew at all the way. For sometime after take off her flight was subsonic until she'd refuelled. There were also considerations relating to minimising the effects of a sonic boom over land. And there was the need for subsonic flight many hundreds of miles out from her landing point. It meant that in about three hours time she would begin her re-entry 400 miles west of Cornwall before passing over Fastnet Rock 35,000 feet below and then along her glide path into Fairford. They had made good time and were a little early. It was 3.45am.

∞

Sir Jack Geisner estimated he was one of a party of thirty or so people, mostly scientists and technicians, who had gathered back in the former aircraft hangar housing the Projector Gantry and Control Room. He was beginning to feel the effects of a long day, and was now running on nervous adrenalin induced by the moment. He heard sounds emanating from the roof fifty feet above. A section was sliding back and as it did, the gantry, in the centre of the floor, began rising upwards and through it. A loudspeaker stuttered a muffled announcement: 'The countdown has begun. All Control Room staff to their positions.'

As he looked, Sir Jack was reminded of the scene at the end of every James Bond film, where the villain, bent on world domination, has begun a countdown that, if left unchecked, would result in certain death and destruction. But here was a difference. In the Bond films, it was always amidst frenetic activity. Here, there was virtual silence and studied calm. There was no flashing light, no warning klaxon, no shouting of orders. It was 3.45 a.m.

A loudspeaker cut across his thoughts followed by a short, palpable silence. And then ...

'Ten!'

'Nine!'

'Eight!'
'Seven!'
'Six!'
'Five!'
'Four!'
'Three!'
'Two!'
'One!'
'We have ignition.'

The disembodied voice tailed away.

Somewhere, far beneath Sir Jack's feet, came a deep rumbling; a vibration he felt through his feet and legs. A brilliant flash from somewhere overhead lit up the entire hangar. The ground moved. The entire gantry seemed to sway.

Absolute silence descended. Even the hum that seemed to pervade everywhere was missing. An alarm went off, followed by another and then another. Some of the people he had been watching a few moments before were clearly worried. Some began to run but in an organised way. There was no panic. The lighting had come back on. Sir Jack began issuing orders. Something had gone wrong, that much was clear. The instant questions were what, where and why.

∞

Fifteen miles off the coast of North Cornwall, Royal Navy ship F83, *HMS Daring*, along with her crew and scientists, had been waiting for the test to begin. It was a few seconds off. The ship's commander knew the essential 'line-of-sight' helicopter was already on-station. After that, he had no idea what to expect. No one had told him. 'That's the idea behind the test,' they had said. 'That's what it's designed to find out.'

All of the crew had been here before on similar tests. Most of them had been to engage in mock conflict with RAF Tornado aircraft, flying up from St Mawgan 130 miles away to the west. For the youngsters on board it was all fun; a time to see if their training really was effective. And sometimes they were actually allowed to fire some of the ship's very expensive and lethal armaments.

The father of the 32-year-old Principal Warfare Officer had also been in the Navy, at the time of the Falklands War. He had tried telling his son about the experience: what really happened when you were hit. He had said never to underestimate the effects of real combat. 'You could never simulate it,' he had said. 'Those times when you knew that the incoming missiles were for real; when you quickly learned the inadequacies of your own defence systems; when you knew the enemy knew all that you knew, because they operated the same defence systems, sold to them by your own government when they were friends and not foes.' His father's words were even now going through his head as he waited the last few seconds.

He wondered too about the latest updates to the ship's defences that they were there to test over the following week. They were to act the role of a 'picket' ship; a sentinel to a theoretical enemy task force. They were to be an early warning and defence line, in case of attack.

He had been fully briefed on his ship's capabilities and had been impressed by its step-change improvement on his dad's old Type 42. But how long would it be before it was sold to a friendly nation who later became an enemy?

He checked his watch. 0345 hours. It was time. He was ready.

He looked at the large circular action data display screen, at all the dots moving across it. One he knew to be the helicopter, but as he watched, the dots with their associated info tags began to hop around the screen quite unlike anything he'd ever seen before. An alarm sounded and the lighting flickered. Simultaneously, he began to feel odd, queasy, as if he didn't belong to his surroundings. One of his fellow officers had ripped his earphones off and was clasping his hands to his head, shouting incoherently.

As he looked around in a semi-befuddled state eerie shapes formed, danced and then coalesced in front of his eyes. Ghost-like figures flitted across his field of vision along with strange sounds. What followed was a brilliant white light that seemed to have been switched on inside his head. He had the feeling of being in a different world from the one he had inhabited a few moments ago. There was something of another age ... of a long, long time ago.

Then it was gone.

The Principal Weapons Officer's vision began to clear. His experience felt as if it had lasted a lifetime but, on checking his watch again, it had been a few seconds. The red emergency lighting flickered on. A few more moments and 'day' lighting was restored. Everything around him seemed back to normal.

He checked the weapon status information screen. His heart leapt.

A missile had gone. He didn't know how. It was impossible.

The PWO sank to his knees with a piercing headache made worse by the dancing images in his head, making it seem as if he were seeing double, with one image being the present and the other being ... being ... he did not know quite what.

CHAPTER 19

It was 6.05 Eastern Daylight Time in Washington DC when the director of the US National Intelligence Directorate, Dexter Coleman, fielded the first of his many telephone calls about AURORA.

He'd been enjoying a deep sleep involving a dream about hunting blue marlin off Cape Cod before realising it was his telephone that was ringing and not the warning bell on the fishing boat. Even after opening his eyes it had still taken a few seconds to realise where he was.

'Thank God,' he said on finding himself at his home in the Georgetown suburb of Washington, his wife still sleeping by his side, and his house not on fire.

The immediate problem was that the call had come through on the house's normal landline, not on his secure red telephone. His first thought was his daughter. Some disaster at university maybe? Possibly run out of cash most likely? He picked up the receiver expecting the worst and was relieved when he realised it was Secretary of the Air Force, Paul Kagan. He sounded kinda excited. Several times he told Kagan to calm down.

'Just received disturbing news,' Kagan said at last. 'Looks like we've lost an AURORA!'

'Shit!' said Coleman involuntarily.

'You can say that again,' said Kagan. 'We don't lose that kinda bird every day. And if we do, the speed and height are so high it's bound to be a total wipeout.'

'Where d'ya get the info?' asked Coleman.

'Straight from Jackson Keyhoe.' Kagan paused before continuing. 'He's Chief Investigator with Air Force Special Investigations Command, Region 7.' He paused for breath again. 'You know how black AURORA's operations are.'

'Sure, sure,' said Coleman.

'It was an AURORA flight out of Kirtland.'

Coleman interrupted him, to warn him about talking on his non-secure domestic line, but the Air Force Secretary ploughed on regardless. Coleman took the risk of letting him.

'AURORA left Kirtland at 1900 hours Mountain Standard Time, due RAF Fairford in the UK at 0400 hours GMT. Apart from her crew of two she was carrying three passengers, one of whom was Energy Secretary Hoop Toberman. However you cut this, Dexter, AURORA is way, way overdue. If she'd come down then it could be anywhere, but mostly over the Atlantic. It's a big ocean. There was nowhere else for a problem to have occurred.'

Kagan stopped to catch his breath before adding, almost as an afterthought, 'News from Region 7 is nearly always bad. It's staffed by guys who look for bad news; whose job is bad news. Wouldn't know a rainbow if it bit 'em in the ass, if you get my meaning. Comes from dealing exclusively with terrorism and counter-intelligence work. But the good news, if there is any, is that unlike any of the other regions, they report directly to me.'

Coleman's mind was racing. Hoop? That really was bad news outta a whole lot of bad news. Hoop had told him about his plan to use AURORA, 'so I can be in two places at once. In the UK and back for our DEEP EARTH meeting with the president.'

Kagan had by now got his second wind and was ready to let it all hang out.

'It smells to me of CIA. One of the passengers was ex-CIA. You know those guys. No one ever leaves. He was on some CIA op, mark my words.' There was a pause before he went on. 'Those sons of bitches might shit on everyone else's pitch, but they certainly ain't gonna shit on mine, not on my watch!'

'I'm making this call to you, Dexter. I wanna see you get a leash on those guys. They're always treading over the flower beds. Someone's gotta get to those assholes,' he said with vitriol. 'Get to Bischoff,' he continued, referring to the long-serving CIA deputy director. 'If he can't keep his dogs on the leash then I'll pass the word along they're running amok. That person will be someone who's close to the president's office, if you get my meaning?' The line went dead. The conversation had lasted little more than five minutes.

Dexter Coleman drew a long, deep breath. He'd spent thirty-one years in politics and knew when a ticking time bomb had been passed his way. A plus point, if there were any plus points, was that it was Region 7 who were involved. So the news could be kept out of the public limelight for a while at least, until they all knew what they were dealing with. But it was only a temporary reprise. Soon after that, there would be procedures and systems to go through, leaving an audit trail visible from the Moon. Sooner or later the truth would finally come out. It always did.

The secure red phone buzzed. Dexter picked it up. It was Kagan again but his voice was now cool, calm and professional. He didn't have all the pieces of the jigsaw, he began, but reaffirmed that it still looked like a CIA mission gone wrong with a 'company' field agent among the missing, '... and whose real role, apart from being a US marshal, is so classified it's way above my pay grade.'

Coleman listened without comment.

'Before I could telephone you again,' Kagan added, 'I had White House Chief of Staff Cliff Wurzburg, wanting to know everything about everything. Now why would that be?'

Coleman found himself murmuring that he had no idea.

'There's been a shoot-out at the O.K. Corral, only this time it's called Trinity Site. Some Iranians involved, I hear, and at the old McDonald ranch, would you believe?' Kagan continued. 'Now why would anyone choose Trinity? It doesn't make sense when they got the whole of New Mexico. Fuck knows what was going on down there, but our guards on sentry duty at Stallion Gate seem to know more than most. Amazingly, a car belonging to one of those involved turned up later on our computer, after it had been checked in at Kirtland.'

'There was only one flight out that night, unscheduled … if you know what I mean.'

'You know who was on board?' Coleman asked, growing increasingly alarmed.

'We know who was on board,' came the response.

Coleman said: 'All I can say is that I don't know too much at this moment.' He paused, collecting his thoughts. 'How long can you sit on this, Paul?'

'Not too long. We have procedures, channels. Then the toothpaste's come outta the tube. Otherwise we get a Congressional Oversight Committee down our necks. I know. I used to be on it.'

'Twenty-four hours?'

'Twenty-four hours at most. And that's only if I stick my neck out.'

'Good. I'll get back to you.' Coleman paused. 'You said you knew who was on the aircraft?'

Kagan had been waiting for the question. 'Yeah. We checked the manifest, and you know that getting anything for an AURORA flight is almost impossible. But we got it,' he said proudly. 'There were three. I got their names. You wanna know them?'

'Go on,' said Coleman.

'It was Secretary of Energy Hoop Toberman, his bag carrier from the Secret Service, and a guy called Tom Santos who I've already spoken about.'

Coleman knew that with the release of these names he was going to have to come clean, or as clean as he could.

'Look Paul, I'm aware of what these guys were doing on AURORA and that it was on the orders of the president. That's as much as I can tell you.

'We're aware too of the shoot-out you referred to, although not any of the details.' Coleman began to get a feeling of nausea in the pit of his stomach.

'What the hell is going on, Dex? Appreciate any further input you can give me,' replied Kagan sympathetically.

'Okay, Paul. Just give me a few hours and I'll get back to you.'

The line went dead. With it came a whole truckload of questions with no answers.

CHAPTER 20

It was 9.15 a.m. on Sunday morning and as promised, Reverend Montanaro had come to pick up Jonathan and his wife in order to drive them both over to attend family Communion at his church in Morwenstow. It was where Jonathan had promised to deliver an inspirational sermon based on the book *The God Particle*. Meanwhile, already in the back of the Reverend Montanaro's red Fiat were the two pensioners he'd also promised to collect from a farm nearby. They too were going to family Communion. It was going to be a tight squeeze fitting everyone in.

Rattling over the green field and unmade road, the car came to rest a few yards from the iron-clad front door of the Lighthouse. It was the first time Reverend Montanaro had come close enough to read a National Trust sign affixed to its four-feet-thick, white-painted wall. The Trust, a national conservation organisation for buildings and land of national interest, had only recently added it to its property portfolio. The sign read:

National Trust
Pendragon Point Lighthouse.
Built in 1896 after the loss of the Indian tea clipper,
The Bay of Panama. In 1937, following severe damage
in winter storms and due to cliff subsidence, it was
rebuilt to house a 570,000-candlepower light visible out
to sea for a distance of 24 miles. Pendragon light was
one of the last manned lighthouses in the British Isles until
replaced by an automatic beacon sited five miles due south
on Puffin Rock.

There was a doorbell. Reverend Montanaro pressed it and waited. As he did, he noted that all around seemed deathly quiet. There were no seabirds as he would have expected and whose cries would normally have woken the dead. He pressed the doorbell again. Still no one came. He tried yet again, and yet again there was no answer. Hesitantly, he tried the door and pushed it. It swung open. He went inside. There was no sound of activity. No one. Nothing.

He was beginning to get a bad feeling. He searched the ground floor apartment and found two bedrooms, one containing what appeared to be a sleeping Dolores and the other a sleeping Ben. He ventured out and then up the iron staircase. It led to a first-floor landing with a single door leading off, followed by a further flight of stairs. He ventured tentatively further upstairs. Maybe Kat and Jonny were in the lamp room where they hadn't heard him. He ventured further up. Still no sound.

'Hello!' he shouted. There was no response. 'Curiouser and curiouser,' he said to himself.

He came to another landing and another flight of stairs. He decided to try this one door. It revealed a bedroom. Inside, he saw Jonathan and Katrina. Katrina was still in bed, Jonathan was lying on the floor in an obviously unnatural position. Neither stirred. His heart leapt. He knew enough about first aid not to move them, apart from checking for a pulse. They were both alive, of that much he was sure. He felt inside his own jacket pocket for his mobile phone.

'Shit! Shit! Shit,' he said aloud in an exasperated voice. 'Out of fucking range, and just when you need it!'

He heard a noise. It was soft and low, and gentle and human, and female, coming from somewhere lower down the lighthouse tower he had passed moments before when he was going up the stairs. He went to investigate further, and to look for a telephone landline on which to make an emergency call.

∞

It was the same dream. It had always been the same dream ever since she and her warlord had watched the strange craft with its even stranger occupants fall from the skies that summer evening many months before. Now the messenger had come from her warlord just as the dream had always foretold he would. The messenger had said her warlord was now in the ancient city of Aquae Sulis a day's ride to the west. Her warlord wanted to see her this one last time. Before he met the Saxons in what may be their final battle.

Although she knew she must go, she dreaded it. For the dream had foretold everything, therefore she knew where and how it must end.

Since travelling eastwards to Ynis Witrin to be safe with Myrradin many months before she had felt a change creeping over her. She had sought solace in Myrradin's company, drawing strength from the last of the Druids, whose abilities came from years of studying the earth around him, from studying the stars, and from the accumulated knowledge of his forebears, who in turn had studied all things and therefore knew all things. 'There will be an inevitable spiritual reincarnation,' he had said. 'There is a guarantee of life eternal that will be granted to you and to your warlord. Do not worry.'

She had told him about her increasingly recurrent dream, how she saw a strange world unlike any she could comprehend. How she saw its mysterious and huge buildings, one white and tall with a beaming light standing at the edge of the world. How she saw strange objects without horses drawing them move back and forth of their own accord over green fields or along metalled tracks. How people dressed in brightly coloured and tightly fitting clothes.

Then her dream would take her to where they had both been so many times before, to where the hot waters bubbled to the surface, where the seven hills met and the river bent. Where she drifted along stone corridors where oil torches cast their flickering shadows. Where in the background came the sound of running water, where there was hot steam, where men and horses gathered and a million things were done as demanded by the coming battle. Soldiers, infantry, cavalry, crowded in on all sides. She could see her warlord talking, giving instructions. Always she tried to reach out to him. She would call his name. He never heard. For in the dream, she remained a ghost; a wraith no one saw or heard.

Then she would move high into the air into the brilliant, almost blinding white light and bitterness of a winter's day. Yet she was cocooned from the cold. She looked down on to a faraway earth, on to the decaying ruin of the Roman city once a crossroads of empire, where once legions, merchants, politicians had discussed the fortunes affecting a million men.

Now it was silent, deserted, the legions long gone to protect Rome leaving the British alone to defend themselves against invading Saxons, Picts, Scotti and Irish. The British tribes had fought them individually until united by her warlord. He alone had given them new fighting skills, new tactics, strategies and wisdom. And in return they had given him their allegiance and their respect. He had brought his Sarmatian-descended warriors, with whom he had won so many victories before and culminating in a golden age of peace. But it had been so many years ago. Now he must fight again, against new Saxon invaders.

She was above the battlefield now. It was not a battlefield of heroic splendour, but one of mud and blood, hacking and squalor; of severed limbs and hideous injuries; of bodies half-covered with the contrasting white of falling snow. She saw him again, as she expected; alone and brave and courageous. She cried out, had tried to reach him, tried to help, tried to hold and to hug him this one last time. But she was a ghost, a prisoner of the same dream that held them both. In a remote way they were both locked in another place and another time. She cried out, she screamed. But there was no one there.

As if lying at the bottom of a deep pool, she struggled and fought for breath. The clammy, warm waters tried to hold her as she suffocated. She fought harder, fighting, swimming, until finally she broke free. She kicked out and broke the cold surface as the last of her air expired. She splashed towards freedom, towards warmth, towards 'the light', and what lay beyond.

Her eyes flickered open. Relief flooded into her soul. She was here, in this world, in this time and place. Her name was still Ganhumara of the dark hair and violet eyes. She breathed a sigh of relief. But despite it, she felt something had been lost that she did not understand, or know its reason. In its place was an inexplicable sadness at what might once have been.

As her rational mind asserted itself all she knew was that it was morning and she had to leave. There was a ferry coming to collect her. She must ready herself.

Two female helpers from the Holy Order helped her dress, and then escorted her to the small boat waiting to take her from the Island of St Brigid, to Ynys Witrin and then on to the shore where the Sarmatians would be waiting to escort her to the city between the hills, where the hot springs flowed.

The boat began pulling away. She looked back at the island and wondered when next she might see its Healing Order of Women, when next she might help in its regime of prayers and penitence and with tending the sick, infirm and dying.

Gradually, with only a light wind filling its sail, the boat drew nearer the island of Ynys Witrin. There she would meet with Myrradin before he escorted her on to the next ferry waiting to take her to the mainland a short distance away.

As her small boat approached Ynis Witrin she remembered the circumstances of her arrival on the island that last summer. How the summer had been drawing to its close. How there had been several badly burned bodies in the strange craft from the skies. Just one had been barely alive. How to ensure his survival her warlord had requested the merchant take him, along with herself, to Ynys

Witrin where the healing powers of Myrradin could be sought. A few imperial gold coins had changed hands before it had been agreed. How they had sedated 'the survivor' for his own safety during the voyage. How she had cut his clothing from him to apply the herbs and special potions she had prepared. She had noted the figure had been large and well built with a close-fitting, light blue material covering his body from neck to foot. There were tight-fitting boots too, of a material whose manufacture was beyond her imagination but shaped like Roman cavalry boots. She had seen that the face was burnt, the hair partially singed.

Then, before she had departed with her patient, Ar-tur had taken from him the large amulets that had been hanging around his neck. 'They may choke him if he is seized by a fit during your voyage,' he had said just before the merchantman's corbita had set sail. 'Keep the amulets safe' Ar-tur had requested. 'One day when we are together again we may cast them into the hot waters of Aquae Sulis. We will do it on his behalf as a token to our God to keep him safe from harm.'

Then she had forgotten their existence until packing a basket she would taking with her. It was almost as if their God had willed it. She must not forget them again she had said to herself.

Then she remembered again the strangeness of the man's clothing. How on the chest, had been displayed two insignias, the meaning of which they had all been unable to decipher. There had been writing too, above the insignia on the right-hand side. It was on a black background with white lettering:

United States Air Force

0VAF105

On the left-hand side, of equal height to the right lettering, was a further, similar sign proclaiming:

Nomex® Limited Flameproof Protection

US Military

Du Pont™

Over the months she had been at Ynis Witrin the man from the stars had made a slow recovery. From appearance he seemed old and wise. But their world continued to appear utterly strange to him. He talked too in a language no one understood whilst at the same time he did not comprehend theirs. But although he had understood nothing to begin with, he learned fast. Now she had had to leave him in the hands of others as she made her journey to the west, to the ancient city of the Romans, Aquae Sulis.

Her boat from St Brigid's reached Ynis Witrin quickly. Myrradin was ready when she arrived. As always he looked the same: inscrutable and ageless, with his long and sparse grey hair, the same threadbare linen outer garment whatever the weather, his seeming indifference to the outside world. Their meeting had been wordless but he had kissed her once lightly on the left cheek. It was enough, for she knew how much it meant. Silently, they had walked down to the jetty together, to where her last ferry would come.

She saw it draw slowly nearer. As she watched, she knew it would be a day's ride from the far side to her warlord's own Sarmatian led army encampment. According to the messenger it was where the seven hills met, where the river curved and flowed in the way of the dragon. The way of the dragon? She had turned the phrase over in her mind. Where had that come from? It was Chinese! But who or what were the Chinese? There was no answer. It remained a mystery.

The ferry bumped against the side of the wooden jetty. Myrradin kissed her once more as a friend does, then made the sign over her head for his gods to go with her.

The oarsman was soon propelling the boat back to where it had just come from. She looked back at the always mysterious figure of Myrradin and wondered just what he knew. The boat rocked a little.

Glast was waiting for her with his five horsemen. For years he had been the warlord's closest advisor. Soon she would be with his men ready to begin her journey westward, towards seeing her warlord. Her heart ached for him, for his arms to be around her, for him to hold her, to feel the beat of his heart next to her pale skin, for them to make love for what she feared would be the last time. She knew that loving someone as much as she loved him meant you prized their happiness more than your own.

Glast said, 'It is a day's ride from here. We must be there before the light fades, taking care not to meet any Saxon raiding parties. We know they are active along the route we will be travelling.' He helped her out of the boat and on to her horse. 'The men are a precaution,' he said. 'But should we meet Saxons our best course will be to outrun them. It is why we have chosen our best horses.'

Soon after their departure it snowed and the earth became blanketed in white, deadening all sound. The hot breaths of their cantering horses condensed in the cold air. Not a bird cried nor a tree rustled. Nothing. Only the leaden skies for company as the snow fell. They rode all day until the light began to fade in the late afternoon. Then far off she saw the city lying across the river where the seven hills met. As they approached they allowed the pace of their horses to fall as the weather worsened. It was then she again recalled the words delivered by the messenger:

'We are camped in and around the Roman city of Aquae Sulis to the west. It is less than a day's ride. A Saxon force is camped to the far south, led by Aelle and his three sons. They are close to the coast where the Roman shoreline stands. More men arrive by the keel-load each day. The fighting will begin soon. Time is short. I must see you maybe for this one last time; before this last battle I must fight.'

The messenger had added where she could find him, where they had been many times before, where the hot springs bubbled to the surface in the bathhouse. It would be the ideal place to make their offering she thought.

The Sarmatians came down the hill to the east of the city, to where the ice-encrusted ford allowed them to cross the swiftly flowing river and where ahead lay the city.

Darkness was almost complete as they passed by the fallen columns and collapsed roofs of long decayed buildings. Soldiers on horseback came out to challenge them but as they too were Sarmatians they recognised those who had accompanied her on her journey from Ynis Witrin. They escorted them into the temple courtyard while one of their number went off to find her warlord and their revered leader. Meanwhile her own escorts dismounted to join other horsemen and soldiers already in the courtyard, some of whom had only arrived earlier that day. Whilst she waited she learned her warlord and his men would be breaking camp early the following morning ready to march on the Saxons, whom they estimated were several thousand strong and less than half a day's march away.

Briefly Ganhumara was left to her own thoughts, unsure where to go or what to do. She saw the fallen Doric columns and the once-high pediment bearing the face of the godess Sulis Minerva lying on the ground partially covered by snow. Rubble lay scattered over the wide flight of steps leading up and into the hall of the bathhouse. She remembered climbing those same steps many times before usually with her warlord. As she looked around, her horse shifted nervously, sensing something new in the atmosphere. Ganhumara wiped snowflakes from her eyes and pondered why she still called him her warlord. Why did she not call him as others called him? Ar-tur, the Great Bear of the Latin language.

She glanced upwards at the heavens. The clouds were clearing, letting a full moon cast an eerie and baleful light that flooded onto the surrounding buildings. In the near distance, across the courtyard, past the fallen statue, she saw the lone horseman approaching. Her heart jumped. She could see him looking at the temple's moonlit façade. She followed his pointing finger to where he goddess's face would once have been.

'Let us hope Sulis Minerva watches over us tonight,' the figure said as he drew alongside. He meant it lightly, for he did not believe in the old way for like her he was a Christian.

With a swift, flowing movement he dismounted, seizing the reins of her horse, steadying it and soothing the animal. He called to one of the men who was passing and beckoned him over. He spoke a few words of command to him before bidding her dismount. As her horse was led away he embraced her tightly while kissing her passionately on the lips. She relaxed into his arms. Then he led her up the steps so that they walked below the temple's colonnaded entrance and into what she knew the way to the bathhouse and its interconnecting rooms.

At the step's summit they halted briefly. Then they continued along a corridor, whose smooth plastered walls ran into the distance, lit at irregular intervals by flickering oil lamps. They walked a few more steps towards the sound of running water. Then Ar-tur stopped briefly and stared about him as if a blind man had suddenly been given the gift of sight but he said nothing. They moved on before finally coming to a room obviously prepared for them both. She felt its warmth and humidity on her cold face as steam wafted unobtrusively in from somewhere As she stood he turned her around so he could embrace her once more. They kissed as if it might be their last. At last they disentangled before subsiding onto a mound of fur skins and woollen blankets forming a low bed in the centre of the room. They were alone. It was what each had been waiting for. There was much to say but so little time in which to say it. For what seemed like an age they said nothing. They let their bodies say more than words ever could. Then they made love.

He knew it was early mornings and he was awaking from a much needed deep sleep. He turned over and saw her in the dim light as if was for the very first time. Yet he knew he had always known her. His arm reached out for her as hers reached out for his. It was instinctive. He wanted to feel her comforting warmth before he must go, before he must leave he for this one last time. They were lying together naked. They had to dress. He told her the way it must be and she seemed to understand as if she had always known how it must end.

Together they walked hand in hand into the stillness of the large bathhouse. As they walked in his mind's eye he could see it all from above, even though he knew it to be impossible. He could see the snow-covered landscape, the surrounding hills. He could feel the cold, see the temple with its fallen walls, its fallen roof; the desolation of centuries of decay.

She spoke. 'There are so many questions,' she said, 'and too few answers. All I know for sure is that we are here, in this place and in this time. We must act together.' She repeated the last few word as if they were especially important. 'We must act together,' she said again before adding. 'There is no other way.'

He knew her words were true, just like they always had been. 'It's you, isn't it? It was always you?' he asked. There was a short silence filled only by the sound of running water coursing its way up from deep underground, then along skilfully built channels, then through lead piping before discharging into the hot pool.

The warlord listened momentarily to the symphony of sound before pulling her closer to him. He could make out her face. He looked down into it and said again, 'It was always you. After all this time you knew from the beginning that it was always me. But you never said. Not even after the eons we have already spent together. You remained silent, leaving me to realise for myself, at this moment.'

'All the time?' she echoed softly. 'All the time,' she repeated. 'It has been but the passing of a moment. I would gladly do it again

and again and again. And I know we must. It is the way of things. I ... I could never say. Never. It was not allowed.'

Tears welled up into the violet irises of her eyes. Something new turned over in her mind. 'All that I ask you to believe is that I came back for you! I volunteered to come back because I loved you,' she said, her voice rising in exasperation. Although he was a big man to her, at that moment he was a small child. 'I know,' she said, her voice trembling. 'I know it's been hard. We cannot know all things. I had to wait. It was hard for me too. I had to be certain. But come. We must move from this chamber to be beside the pool'.

For a moment he was angry, but the future was beginning to tumble back into his once-locked memory.

She saw the anger pass. 'Do you forgive me?' she asked with that serene tenderness that had captured him from that first moment that was no longer in the past, but in a distant future he had already shared with her.

She pulled him closer. 'Soon, very soon you must go,' she said. 'We will be parted for this one last time.' Then she added: 'From then on we will be given eternity.'

'How do you know?'

'I know,' she answered enigmatically. 'Is not your God, our God, the God of love?' She continued, 'Did He not say that of all commandments none is greater than this?'

'Yes, yes ...,' he muttered softly, only dimly comprehending.

He placed his arms to enclose her and to hold her tightly, kissing her with the longing tenderness passed down from the far future. She let him but then had to be free. She said as she stood beside him, 'I have to give you these.'

From beneath her white Roman stolla she withdrew the metal objects the ones retrieved from the 'man who had fallen from the sky'.

She placed them in his hand. They felt warm, alive. Her warlord looked at them for a few seconds before he remembered.

Ganhumara saw that look on his face. 'It is mostly for the man from the sky. But he will not mind it if it is also for us. You must throw them into the waters in the same way the Romans once did, as was their custom and those before them. We are Romans just as we are Celts. We are in their temple. It is our temple. To seal the future we must do the same. It will protect us... It will protect us,' she repeated softly, her hands enclosing his hands that now held the discs.

'The waters of the gods will protect them for all time, until it is time...the right time.' She removed her hands from his, urging him on. 'As they enter the water we must both make our wish.'

He looked at her, at the hot pool at their feet, then at the objects in his hands. He saw the glinting of the many coins and jewellery already thrown there by generations who had gone before, as they too had sought something from a future still to be written.

The warlord, the man called Ar-tur, the same man as the man from the future, looked at the objects, the amulets, he held in his hand for the last time.

In the flickering flames of the torches he could see their inscription but did not understand their meaning. The inscriptions were identical on both amulets he noted. They read:

Property of the United States Government.
If found return to Los Alamos National Laboratory.
Shipping address: PO Box 1633, NM 87545 USA.
Serial No: 175890/1603/11.
Possession is an offence without proper authorization.

Ganhumara looked at her warlord. Time was short. Hope was all she had. And love. Just as it had always been, she knew it would never change. They were now locked into the endless loop of time where there was no beginning and no end.

The warlord drew his arm back. He turned to her with a puzzled look on his grizzled face. 'I feel we are being watched by an unseen presence.'

'It has been the same for me since leaving Ynys Witrin.'

He did not reply. With a shout, he drew his arm back, then brought it swiftly forward, releasing his hold on the amulets. He and his queen made their wish.

The amulets curved high into the air to be swallowed by the darkness, before descending and splashing, finally, into the steaming waters of the temple of Aquae Sulis.

They both watched for a moment, then turned and walked back along the long corridor and out into an approaching and cold morning with its sky twinkling from a billion stars. As they watched a ripple seemed to cross the heavens. With it they were both plunged into an infinite void.

∞

Pendragon Point car park had become host to a surprising number of vehicles for the time of year, including two police squad cars, a specialist incident unit, an ambulance whose siren had been silenced but whose blue light was still revolving, an unmarked police detective's car, and an unmarked car belonging to someone from British military intelligence. Yellow incident tape had been used to mark off a large area surrounding the lighthouse. Men were going in and out of the building, either carrying equipment into it or bringing household items out for later examination.

Two ambulance paramedics appeared out of the doorway carrying a stretcher on which lay a semi-conscious and moaning Dolores. A few moments earlier they had brought out what appeared to be a semi-conscious Ben, in his wheelchair, before loading him into the back of the ambulance.

A uniformed constable had been stationed at the empty wooden hut normally reserved for a car park attendant during the summer months. Within a few minutes the ambulance became the first vehicle to depart. As the constable continued gazing in the direction of the lighthouse a bright red Cornwall Air Ambulance EC135 Eurocopter came into view, and descended slowly into the middle of the car park. With its main rotor continuing to turn, two green-clad paramedics emerged pulling a lightweight, collapsible patient trolley behind them as they ran in the direction of the lighthouse and disappeared inside. A few moments later, they reappeared with a woman strapped to their patient trolley, with an intravenous drip supplied by an attached fluid bottle. Within seconds they reappeared and were dashing back into the building. The constable watched with mild interest. Again it was not long before the paramedics reappeared, this time with a man strapped to the trolley. He was loaded on board, the two paramedics jumped back into the helicopter, and with a roar it lifted off leaving the constable to wonder which hospital it might be heading for, and pondering only that it had to be serious to have involved the air ambulance.

With its departure, silence settled on the area once more. Gradually, and one by one, all the vehicles made their departure except the one belonging to the incident unit.

'Stay until they go,' the police constable in the wooden hut had been told. 'Could be a couple of hours yet.'

No one left and no one came during the following hour. To relieve the tedium he pulled from his coat pocket that morning's edition of *The Guardian* national newspaper. One of his colleagues at the police station had handed it to him after finishing reading it himself, with the words: 'Read page six. Not often this town makes it into the national press!'

He quickly located the story. The headline ran:

CORNWALL HIT BY EARTHQUAKE

before following with:

The 2.2-magnitude quake hit Bodmin at 2.40 a.m. on Sunday and lasted just a few seconds, the British Geological Survey (BGS) has said.

There were no reports of damage, but the tremor was felt in Bodmin, Liskeard, St Austell, Padstow, Camborne, Wadebridge and Callington, and as far away as Bude.

Anne Cairstairs, from Bude, told the BBC: 'It was very loud, like a whooshing, and the house and bed were shaking. I thought I'd imagined it.'

A number of earthquakes are felt every year in the UK, but most are very small and cause no damage, the BGS said.

The most powerful earthquake recorded in the UK occurred off Dogger Bank in the North Sea on 7 June, 1931. It measured 6.1 on the Richter scale.

CHAPTER 21

'You all know what's happened. The question is where do we go from here?' The president looked grim. 'Just to make sure we're all reading from the same page, Dexter here will tell you what we already know.' The president paused. 'Over to you Dex, and make it quick. There's things to be done, even more than before.'

Besides the president and Dexter Coleman, there were three others who had been summoned to the emergency meeting in the Roosevelt Room: White House Chief of Staff Cliff Wurzburg; Chairman of the Joint Chiefs Admiral Nelson W Loadhammer; and Deputy Secretary for Energy Howard Davenport. Although it had been Cliff Wurzburg who'd done the actual summoning to the meeting, everyone accepted it was Coleman who, by virtue of his office, was more ahead of developments than anyone else.

'First, is what we know for sure,' said Coleman slowly and deliberately. He paused. 'We've lost an AURORA, number 006. The last one built. We lost contact with her at around 0345 hours GMT or 10.45 p.m. Eastern Standard Time. As far as we know it was somewhere over the North Atlantic, as she was closing in on the UK, prior to making a landing at RAF Fairford near Gloucester. She was to be turned round and sent back the same day with a scheduled landing well out of sight at Andrews. We already had a C130J module parked at Fairford to greet her and turn her around ready for the flight back.

'At the height and speed she flies it's not surprising that if something goes wrong, it's likely to be "lost with all hands".' He paused to let the importance and implications sink in.

'Apart from the crew of two, we know she had Hoop on board. We assume he had with him his Secret Service officer. I already have the aircraft manifest, but we know there was a last-minute change where one of my guys from the NID was seconded at the last minute to act as "security", as a shotgun messenger if you like, for Hoop while he was outta the country. This because his normal Secret Service officer in an emergency would already have his hands full carrying Hoop's "football". We normally have it covered but this time around his "other" Secret Service guy went down with what we think was "flu".' Coleman pulled a face, shrugged his shoulders and added, 'It happens! Even in this man's army. So we live with it!'

'Ironically, Hoop was on board because he was trying to make it to a meeting he had in the UK, and then to get back here in time for a scheduled president's DEEP EARTH meeting the following day.

'Our NID operative was a Tom Santos. He stepped in at the last moment to act as an escort for Hoop, who was flying out from his patch of territory in New Mexico. It seemed like a good fit.

'What we didn't know at the time was that by sheer bad luck, we'd end up with Hoop's DEEP EARTHcomputer system on board AURORA, together with those Tom Santos ended up carrying. On top of that we've got the two security "Code Black", flash hard discs, used by our Iranian professor to store data, and as an encryption key to gain access to his partition of the DEEP EARTH computer partition at Los Alamos and at Berkeley.

'In short, unwittingly we've scored a triple whammy. As you all now know, we had last-minute trouble with the professor. It resulted in our agent temporarily taking charge of the classified computer equipment the professor had been using, to forecast what tunnels we should dig, and where, at Hanford. Those last-minute calculations were necessary to tell us where and how to bury our radioactive past, together with a nuclear device whose detonation would bury the whole lot.

This, so that when the predicted earthquake hits us as forecast, all we'd have to worry about was the normal destruction: buildings, structures, what you'd expect. But what we wouldn't have, if everything had gone to plan, were lethal amounts of radioactive waste washing down the Columbia River and contaminating the area for the next 100,000 years.

'This separation should have meant we wouldn't have to evacuate most of the entire states of Washington and Oregon on a more or less permanent basis. If DEEP EARTH doesn't go off as planned then we estimate several million Americans are gonna be dead or homeless. Possibly both.

'So that gives us at least a million reasons to try and fix the mess we've got ourselves into. That boils down to locating the Iranian professor's computer and storage discs pretty damn quick. Of these, the most important are the smallest pieces: the two "Code Black" computer discs. Ideally, without letting the British know about the enormity of our problem. To them we're gonna keep playing the card that it's our top-secret security equipment we're most interested in, plus the not-small fact that AURORA herself was classified, which all has the not-small advantage of being true.'

'Go on, Dex, tell everyone the kicker,' said Wurzburg in a quiet voice.

'Don't ask me to explain this,' said Dexter. 'It's just as we've been given it by the British. Evidently, they have wreckage belonging to AURORA and it's pretty much in one piece. They say it's the escape module. They also have a number of artefacts found with it, including clothing, personal effects, and personal computer equipment. So we could be in luck. Sadly, they haven't mentioned anything about bodies. The question here is how come they have personal artefacts and no bodies? It doesn't make sense. Also they've been suspiciously quick off the mark in locating the AURORA wreckage.' Dexter held up his hand. 'Don't ask me any more. What I've said is as much as I know. But from what we've got, we've not been dealt out of the game. We could be in luck.'

The Chairman of the Joint Chiefs, Nelson Loadhammer, was going to say something when he was stopped by a gesture from the president.

'Thanks, Dexter,' said the president. He looked around the table as if daring anyone else to speak, before adding: 'There's a whole lot more we don't yet know. Not least of which is that two of our most important people connected with DEEP EARTH are also connected, one way or another, with this tragedy. One, Hoop, we can presume is dead. You'll have to take his place at short notice, Howard. The other is this guy Santos, who we can also presume is dead. He gets himself involved in a shoot-out at the O.K. Corral down New Mexico way, with two Iranians, and involving our own DEEP EARTH expert. With both of them gone, I think we can safely say we're gonna have trouble in the near future finding out who did what to whom in New Mexico, and why. The professor is in a cell in New Mexico, one Iranian is dead and the other in a hospital in Albuquerque.

'But there is a chance, as I understand it, that we might recover the lost data and encryption key, and with it save several million Americans from either dying or being evacuated.

'One thing is certain, we're gonna have to get them out before the quake strikes, and we end up with more radioactive waste than we can cope with.' He stopped and looked around the table. 'Gentlemen, what we don't have is a lot of time. Today is December 30th. So I'm gonna give it until the middle of next month, January 14th, before I ask Admiral Loadhammer to press the button, and speed up our alternative evacuation plans. I don't have to tell any of you that these will make Hurricane Katrina and New Orleans look like a tea party by comparison.'

There was silence from everyone. 'That's sixteen days' time. All I've gotta add is that I'm asking Dex here to get on the next plane to London, to find out what he can ... and, more importantly, to do what he can.'

With that the president adjourned the meeting. As everyone got up to leave, Admiral Loadhammer said: 'Just wanted you to know that an aircraft investigation involving a US military aircraft will fall under the jurisdiction of the Third Air Force based in Germany. They'll already have a team on it ...'

'They do indeed, Admiral,' said Coleman, 'but because of the sensitivity affecting this particular accident, we're asking them to hold off until I get there.'

'Anything you want me to do?' asked Loadhammer.

'Nope, Admiral. We've got everything in hand.'

'Okay. You know where to find me.' Then, almost as an afterthought, he added: 'Don't think I'm being facetious here, but you will know what you're looking for when you get to London?'

'No, I don't mind the obvious, Admiral.' As Coleman was saying it he reached into his open briefcase and pulled out a thin file. It contained a number of foolscap-sized photographs of people who the Admiral presumed were AURORA's passengers. Coleman slid one of the photographs across to him. Two circular objects were photographed against a one-cent coin to give an idea of scale. One of the objects had a blank face, the other was inscribed:

Property of the United States Government.
If found return to Los Alamos National Laboratory.
Shipping address: PO Box 1633, NM 87545 USA.
Serial No: 175890/1603/11.
Possession is an offence without proper authorization.

Admiral Loadhammer said nothing, and slid the photograph back to Coleman, who replaced it silently in his briefcase.

Everyone began moving away from their seats and heading for the door. As Coleman was preparing to leave Wurzburg stopped him for a short, private conversation. He waited for the room to empty before he said in a conspiratorial voice: 'Be careful with the Limeys. It musta crossed your mind that something about this whole AURORA thing doesn't smell right. Our friends over the water can be sharper than a wagonload of monkeys. So be extra careful. Sure, we need their help and as much as they can give, but in exchange we don't want 'em sticking their nose into our domestic business. We can make a big mess of it on our own without their help!'

Wurzburg opened the door and they both walked through it. Coleman turned right and walked down the corridor, while Wurzburg walked ahead, opening the door of the Oval Office. Wurzburg's voice sounded down the corridor after Coleman. 'Your flight ticket and details will be fixed by the time you get to Dulles … and if we have to hold the plane 'til you get there, then trust me, we will! So don't kill yourself – or let your government chauffeur kill you – speeding. You'll have police outriders clearing the road ahead of traffic, so don't even think about it.'

∞

It had been snowing for three hours when Dexter Coleman's limousine pulled out of the West Wing car park and headed on to Pennsylvania Avenue. Two police outriders joined it. They crossed the Potomac using the Roosevelt Bridge before heading on to Interstate 66. His driver had already collected his emergency overnight bag from his Georgetown home and kept ready for just such an emergency as this. It was an old habit he still kept and a left over from his field agent days. He silently thanked God it was all happening after Christmas, when he'd already done his duty as father to his two teenage daughters.

Despite the snow, Dulles was an 'all-weather' facility so the Metropolitan Washington Airport Authority had kept all eight runways clear. It meant that British Airways flight BA0264, cleared for take-off at 8.24 p.m., was still on schedule. More importantly for Dexter Coleman, it meant the 192-seat Boeing 767-300 would still arrive in London seven hours and twenty minutes later.

Commercial flights out of Washington left late in the day to give passengers an option of catching at least some sleep before arriving in London, ready for business. Coleman had, of course, considered using the NID's own Gulfstream G550 twin-engined passenger jet, but the advantage still lay with the commercial option, after taking into account how long it would take to get the Directorate's own aircraft ready to fly, along with rounding up a crew.

Meanwhile, the White House had already notified the US embassy in London's Grosvenor Square to have a car waiting, ready to meet him shortly after touchdown at 0845 hours GMT. He would, of course, be cleared on arrival at Heathrow, through normal diplomatic channels. It all meant Coleman would hit the ground running, a vital prerequisite considering the stiff time pressure under which he'd be operating.

As his car sped over the Roosevelt Bridge heading out of Washington he ironically wished AURORA flights had not been suspended because of the very accident he was investigating. However you cut the cookie, it was still the only alternative transport that could have reduced his travel time significantly.

Coleman's car soon became snarled up in traffic as it sped along Interstate 66, and was soon reduced to a crawl. Coleman began formulating a plan of what he would do on arrival in London.

His first action would be to head for Thames House in Millbank for a meeting with the head of MI5. They dealt with all matters relating to internal security under which the missing aircraft would most likely fall, particularly as AURORA was a military matter. Coleman had known the current and long-serving 'M' for fifteen years, after operating successfully together in their younger days on several overseas operations. As a result, they'd developed a level of mutual respect and trust for one another. It was just as well, he thought, for the question of AURORA might require both.

Despite the traffic it was forty-five minutes later that Coleman appreciated being the recipient of the legendary White House organisational skills. It came as he sat down in a first-class window/recliner aboard Flight BA0264.

Ten minutes later, he heard the twin Rolls Royce 211-524G power plants spin up as the 767-300, weighing nearly 184,000 kilograms fully loaded, hurled itself down Dulles Airport's runway 01. At 145 knots the nose wheel lifted off and at 175 knots she lifted clear. Speedbird 264, as she was identified by Air Traffic Control, climbed steadily to 3,000 feet to become one of 1,000 similarly configured aircraft regularly flying the North Atlantic route. Coleman knew that flying east meant flying with the prevailing Jetstream and it would therefore reduce flight time by almost an hour.

∞

Thirty-nine thousand feet below and nearly four thousand miles east in London, Sir Freddie Stirling's black Jaguar edged its way along London's Whitehall before turning into Horse Guard's Parade. The car stopped outside the white and gold façade of the Horse Guard's Hotel, where Prime Minister Gerald Campbell was handing out cheques to winners of the annual Young Entrepreneur of the Year competition run in conjunction with a national newspaper.

All Sir Freddie currently cared about was that just when he needed a considered chat with the PM, the best he could get was probably going to be hurried exchanges in a hotel lobby or corridor, before the PM was whisked away to the Commons to attend 'PM's Question Time'.

Sir Freddie was escorted into the hotel lobby followed by his armed specialist protection officer. The PM's wily and blunt director of communications was there to greet him. Sir Freddie was shown immediately into a private room. He was thankful for the small mercy that he wasn't going to be discussing state secrets in hotel corridors or subterranean passages, as he'd been thinking a few moments before.

He'd been in the room for less than a minute when the door through which he'd just come burst open and a harassed, but still boyish-looking and fresh-faced, PM entered.

He walked over and shook hands. 'Sorry about all this.' He spread his hands and arms to indicate the inadequacy of the space surrounding them and his own shortage of time. 'Best I could do in the circumstances.' There was nowhere to sit and no table. The semi-darkened room looked more suited to holding conferences, what with its stacks of chairs and tables piled against the walls. The air-conditioning too was obviously off, as the temperature seemed pretty close to freezing.

'I asked to see you because I've had the US president on the phone,' continued the PM, unfazed. 'Thank God you've just been to GCHQ Bude for your first BRIGHTSTAR briefing. Could be a lucky coincidence in the circumstances.'

'Let me get straight to the point. There may be a whole storm blowing our way in the wake of the BRIGHTSTAR mess. Early days for us, I know. But just when we don't want it, we've got some American from their National Intelligence Directorate heading for us. The president has been giving me the heads-up. Naturally, I've told him we'll extend as much help as necessary. That means you, as head of the Joint Intelligence Committee. So I want you to liaise with this guy when he arrives tomorrow morning around 8.45. Do everything short of being helpful. The BRIGHTSTAR problem occurring when it did, and their precious aircraft disappearing right on top, looks too much of a coincidence … even to me. And I'm not there, and not an expert either. God knows what it's going to look like to him. So, Freddie, meet him, keep on top of the situation and, most of all, keep me informed. I don't want nasty questions coming from the president which I can't answer.' He stopped talking and stared straight into Sir Freddie's eyes. 'Have I made myself quite clear?'

'Perfectly,' Sir Freddie answered back in a flat, unequivocal tone of voice. 'Do we know his name?'

'Coleman. Dexter Coleman. Career intelligence. M15 are expecting you for a background briefing. M himself knows Coleman. They have history together, in their younger days. He's good, M tells me. He'll want to get to the bottom of their missing plane quickly. He's no fool, so watch yourself.'

'Evidently important stuff was on board. Usual "vital to national security" stuff, and all that balls. So, like it or not, you're going to be meeting him tomorrow, as soon as he steps off the plane. Liaise with their embassy in Grosvenor Square. Make sure we're not stepping on anyone's toes there.

'We've got to be very careful how we play this. Don't think the Yanks would take kindly to any suggestion we might have accidentally shot down one of their top-secret planes, with the loss of everyone on board. So, let me repeat, be as cooperative as possible … short of being helpful.'

Before he could say any more the door of the empty hotel room opened and the head of the PM's director of communications appeared: 'We've gotta go. You're due at the House in ten minutes.'

The PM turned to Sir Freddie. 'You see how it is?' The PM shrugged before turning on his heel and disappearing from view, and shouting back at Sir Freddie, 'I think we're done,' as he followed the communications director out.

Sir Freddie reflected momentarily on the fact that although a potential bombshell in Anglo-US relations had just dropped into his lap, the meeting had lasted less than four minutes. Was this how all democracies worked, he asked himself?

But there was no more time to ponder the question, for the door opened again and the communications director came back in to tell him journalists covering the PM's recent award ceremony would soon be leaving, and if he wanted to avoid them, then now was a good time to go. Sir Freddie needed no further bidding. There was much to do in the hours ahead.

∞

Dexter Coleman was just finishing his mint chocolate dessert, with a half-drunk glass of passable Merlot red wine on the tray in front of him. He was anticipating difficult times ahead, mostly from his own side. The nub of the problem was finding out as much as he could while, at the same time, keeping the lid on events. According to procedure, as Loadhammer had already advised at their White House meeting, the US Third Air Force (UK) headquartered at RAF Mildenhall would already be involved. Following defence cutbacks, they'd become a nickel-and-dime operation, reporting to Third Air Force (Europe) based at Ramstein in Germany. Mildenhall no longer had the technical capability or manpower for investigating a downed aircraft as complex as AURORA. That would now have to come from Germany. It could be a blessing, Coleman thought. Military aircraft were subject to an immediate news blackout. It was mandatory. The bad news was that it would only be for so long. Because AURORA was military it meant sooner or later it would become entangled with a congressional oversight committee, who would demand to know everything, and more. God only knew where that would end, maybe unravelling all the way to DEEP EARTH.

But that was tomorrow's problem. His immediate task still started and ended with getting back the lost Code Black discs, along with maybe Pahlavi's laptop. Maintaining a security blackout was the next major issue if they were to stop panic spreading across America's north-western states.

∞

Sir Jack Geisner was far from happy as he sat at his office desk in the late evening. He'd just got a preliminary report from his BRIGHTSTAR team on what they thought had happened. He'd scanned it and it hadn't made for appetising reading, although much of it, he'd convinced himself, was sheer bad luck. How could anyone have predicted an earth tremor? Robert Carr had appended his comments too, and they had added a mind-boggling dimension to it all – if it was true.

He hadn't had time to digest it yet. But one thing he had digested was that it could mean closure of the BRIGHTSTAR project on which he'd spent the last four years of his life. Tough questions were going be asked. If there was a government inquiry, as seemed inevitable, the most likely result would be for the guilty to go free, and the innocent to be hanged. Sir Jack already knew where he would need to be. It was an unappetising thought. If he didn't tread carefully it could be the end of his career. Financially, it didn't matter. He'd already earned more than enough to see him out. It was his pride that would be dented, along with a reputation it had taken him a lifetime to build. Soon, probably very soon, the vultures were going to descend.

∞

Sir Freddie Stirling waited patiently, close to the diplomatic gate in the arrivals lounge of Heathrow's Terminal 5. It was

08.40 a.m. With him was the American deputy ambassador. And Flight BA0264 was on time.

While Sir Freddie waited his mind kept pondering the same conundrum he'd been given the day before. For the last twenty-four hours it had kept niggling away at the back of his mind. 'How did she know? How could someone called Katrina Anderson know?' He had asked himself the same question over and over again. There was no answer.

'How could she possibly know before the event, before the disappearance of AURORA itself? How did she know what was inside? The items of clothing? The hat? For it to be possible, all the known laws of physics would have to be broken, let alone the laws of causality. And somehow, for the aircraft to have been found where it was, beneath an archaeological excavation, it had to have come down 1,500 years before it took off. That too was impossible. Finally, was the issue of Dr Jonathan Anderson, their son and his nurse. They were all still in a coma after being airlifted to the Atkinson Morley Hospital in Wimbledon. He'd noted from the report on them that both the son and his nurse had been semi-conscious when they had been found, but had relapsed into unconsciousness soon afterwards. It had to be more than mere coincidence that they all happened to be in the same location, near enough in the beam's path, at the same time as the BRIGHTSTAR test. Sir Freddie checked his watch. It was slightly after 8.40 a.m. and it was December 30th. Where was Coleman?

As he was thinking Sir Freddie saw the tall figure, escorted by two others, heading towards him. He recognised Dexter Coleman from the MI5 photographs he'd been shown. One of the figures, probably the one talking to Coleman, would be from the embassy diplomatic staff, with the other probably from security, Sir Freddie assumed. Within a second or two of seeing him, the deputy ambassador moved forward, his right hand extended in greeting.

For the drive into London it had been agreed, subject to Coleman's approval, that Sir Freddie would accompany him in the back of the ambassador's car as it drove to the American embassy. This was so they could speak promptly and the journey time would not be wasted. They had already exchanged initial pleasantries in the short walk from the arrivals lounge to the waiting car.

During the drive Coleman was the first to speak. His voice was low and soft. Sir Freddie thought it had a New England tinge, probably Massachusetts.

'Let me be frank with you,' said Coleman. 'Sure we want to know the whats and whys of AURORA, not to mention the loss of life of those on board. That goes without saying. We've lost an important aircraft, with important people carrying important equipment vital to our national interests. However tragic it is, the people who were on board, and who we presume are dead, are going to remain dead for a long time.

'But to be hard-nosed about it, what's more important to us at this time is what they had with them.' He saw the concerned look flit across Sir Freddie's face.

'Naw, relax. It's not got anything to do with nukes or defence-related matters. If it was, the Sixth Fleet would be sitting on your doorstep, assuming we knew where to look ... which we don't. So rest easy.

'It has to do with one of our guys on board having an encrypted computer file that's kinda important to us. The file is unique; no copies, no nothing. And before you think it was careless of us, we did have a back-up plan to cover such contingencies. But as Murphy's Law has it, by accident that too happened to be on AURORA.

'So we've lost both, at the same time. Just so you know how it happened it was because an unscheduled passenger unexpectedly happened to be carrying a copy of the encrypted file. It was Sod's Law. He then joined our energy secretary. He believed he was acting in our national interest at the time, which he was, and had no reason to know his action would place that same national interest in jeopardy.

'So there you have it. We need to get the "lost" equipment back as soon as humanly possible. In case you come across them, of most interest to us are two computer discs.' Coleman half- smiled at the possibility. 'They're very small, grey, metal-encased, unbreakable, waterproof, and have the wording "Property of Los Alamos" engraved on them … or something like it.'

CHAPTER 22

Dexter Coleman stared into the deep trench oblivious to the mud, cold and damp of his Cornish surroundings, and not believing the evidence of his own eyes. He was cold and New Year's Eve and Day had been lousy without his family, although the US ambassador had done as much as he could to liven things up. Coleman had been in the UK three days sifting through what the British had told him, documents they'd provided him with, the last voice recordings from the aircraft as she'd approached British Military Air Traffic Control. Then there had been liaising with US Air Force personnel in Germany, more to ensure they didn't get in the way of his own investigations. There were a million other things too, most of which he'd forgotten.

Then the British had suggested visiting the 'accident site' and even arranged transport to it. He'd told Cliff Wurzburg in a short and non-secure call from his mobile that it had meant the so-called 'special relationship' was still alive and active. He was still aware that time was going by and he still had no good news to report back to the White House. Wurzburg reminded him that unless he could report back that he'd found the Code Black discs then everything else was conversation.

The two air accident investigators sent over from Third Air Force in Ramstein were standing with Coleman staring into the trench. Both had remained silent. Then one clambered down into it, followed closely by the other. Although they all talked easily to one another, Coleman got the feeling they were treating him mostly as a detached and remote 'observer'.

Quickly, the investigators began their detailed visual examination of the relatively little that was exposed. Dexter's first thought was that it must be some sort of hoax, but he dismissed it almost as soon as the thought entered his head.

Who would have done it, and why? Maybe it was an elaborate, optical trick where the evidence of his own eyes was designed to miss a deeper reality? That was a nuts thought too. The MI5 briefing during the hour-long helicopter flight from London had not prepared him for what he was seeing. It was simply impossible. It defied all rational explanation.

The focus of everyone's attention lay mostly covered by brown clay. He was still dressed in his long, black overcoat, with green wellington boots, the latter supplied by the British Army. The British had already said they had exposed more of the 'object' since it had been discovered, simply to ascertain what it was; that it was not some student prank but was instead of genuine, military origin. The British had told him that their first thought was that it had been an old World War Two mine, or possibly an aircraft 'lost' from the old Coastal Command base of RAF Cleve during the Second World War.

'Where the base at Morwenstow is now,' they had said, 'it used to be a Coastal Command station. Four-engined Liberators would set out from there on anti-submarine patrol for hours on end over the Atlantic. It could have been an aircraft that didn't quite make it back. But when we saw it we immediately knew we were wrong.'

They'd been careful, they had said, understanding that US military accident investigators would want to routinely examine the immediate area where the object had been found. This was in case it held clues as to why the object had apparently fallen out of the sky. They had also warned of unpredictable high tides at that time of year that could flood the whole site.

Coleman watched the two investigators go about their minute but preliminary examination in their bright orange hi-vis jackets with the words 'Air Accident' emblazoned on their backs. Each was taking photographs, and making copious notes and sketches on their Apple iPads. As he watched, there was not much for him to do except wait. He wanted their thoughts as soon as possible. Time was of the essence. They had said they might be able to provide their findings by the end of the day. Coleman hoped that by simply being there his presence would sweat them into keeping their promise more than mere words could. Air accident investigators were slow and methodical people as they went about their business. They were also part of a military bureaucracy that did not like being hurried.

However, he remained realistic. The fate of AURORA and its whys and wherefores, although mysterious – and he wanted to know the answers – was only essential if it might lead to the recovery of the two Code Black discs. If not, then AURORA could remain someone else's problem, and he would be happy to walk away. After all, it was not part of his daily remit.

The object he was gazing at was large and scorched, of a cylindrical shape, but showing that one end, still obscured from sight, might run more to a point. Lit as it was, by a number of powerful lamps fed from a distant generator, he could see its general outline, together with some of the detail on its matt-black surface. He shifted his gaze from the bottom of the trench to what lay on the three upper terraces surrounding it. Each terrace held a large number of items, all of which had been carefully tagged as to their identity. He could see that closest to him were what looked like corroded swords and spears, intermingled with smaller objects that could have been items of jewellery, such as brooches and armbands. Further away were the red shards of what could have been broken pottery. There were some still only partially excavated human remains. It was all at least 1,500 years old, he'd been told, and of Romano-Celtic origins. Most likely the site had belonged to a wizard, someone skilled in the knowledge of metal working and its use in forging weapons. It was not an unusual find. In that part of Cornwall there were hundreds, possibly thousands, of similar smelters, most usually involved in tin smelting for which the area had once been famous.

A fine drizzle began to fall as Coleman pondered on the artefacts that lay before him. Then there were the other people: British Army personnel, US Army crash recovery experts, intelligence people, two from the Exeter University team who had first discovered the object. They were all holding back, waiting for him to do or say something. For the moment, they could wait. This was now his show. Welcome to Pendragon Point, Coleman thought to himself. Welcome to the Court of King Arthur.

Silently, he let it all sink in. As if to reassure himself he turned to look out over the sea, to the distant cliffs and white, golf-ball- shaped radomes he knew to be one of the UK government's communications listening stations. He resumed staring into the trench, still hoping for ideas or inspiration as to how the aircraft had got there. But none came.

It was made worse knowing that even if the mystery that he now saw was all there was, it would have been enough. But there was more. The helpful Brits had tried to prepare him during the helicopter flight from London. They had tried hard to brief him on what to expect, but what they had said seemed more centred on history and legends: the background of the area. They'd shown him large photographs of the recovery site, along with those of the area in general. Beyond that there wasn't much. 'Early days,' they had said. 'Crash sites take time to investigate.' Coleman reflected later that maybe what he was being told was all the Brits currently knew. Maybe it was relevant and maybe it wasn't. At this stage of the game something, anything, was better than nothing.

So he continued staring into the trench, recalling as he did what his MI5 advisor had told him on the flight down.

Perhaps it was because his advisor had an MA in ancient history that he'd seemed to centre more on the 'colour' of the area, its myths and legends. He'd said legends were often rooted in fact. He'd asked Coleman if he'd heard of William Blake's poem and hymn 'Jerusalem'.

'Yes,' Coleman had said, 'it's an unofficial British national anthem, along with Elgar's "Land of Hope and Glory".'

His advisor had seemed pleased with this response, for he'd gone on to say that the lines in 'Jerusalem' mentioning 'dark, satanic mills' had not referred to north of England cotton mills of the nineteenth-century, as many thought.

'It had more to do with the skyline of 1,500 years ago that must have been peppered with black smoke coming from hundreds of tin smelters.' For added impetus he'd thrown in that its author had also been a modern Druid of the nineteenth century.

Coleman had heard of Druids. An image of pointy-hatted people in long white robes similar to the Ku Klux Klan flashed into his mind. They'd sounded not unlike America's own Anasazi Indians, a mysterious tribe of Pueblo Indians living in Arizona, New Mexico and Utah that had lived for two thousand years before mysteriously disappearing. But what did it mean, other than that there were long-lost and mysterious races on both sides of the Atlantic?

His MI5 advisor had gone on to talk of the Arthurian legend, whereby the king's body had been taken away in a black barge after his final battle at Camlan, and where Excalibur had been returned to the Lady of the Lake. He'd been asked to look out of one of the helicopter's windows at the patchwork of fields, hills and small buildings, sliding past 1,500 feet below.

'Down there is Glastonbury,' his advisor had said, jabbing an index finger towards a small town with a large, conically shaped hill jutting above an otherwise featureless, flat plain.

'Some think that's where Avalon was, where King Arthur was finally buried. Good for the local tourist industry, of course. In Arthur's day it would have been called Ynys Witrin, and would have been an island set in a shallow and stagnant sea. Although drained in the twelfth century, it still floods, even today.'

He had then added that they were flying along a ley line. When he'd asked what those were, his MI5 advisor had said they were straight lines of force the ancients believed linked religious centres, such as Glastonbury, with Stonehenge and even with Morwenstow, where they were now heading.

Coleman's thoughts returned temporarily to the present as he saw first one investigator, then the other, disappear from view as they entered the module's interior.

As there was nothing else to occupy him he tried to recall more of what his learned MI5 advisor had told him: of how there had been many accidents with those that had operated metal smelters; the area had well-developed links with the Middle East, the cradle of medicine; many of their potions had been brought back, and the woman from the local holy order had used them to administer to the sick. There was evidence too of Jewish settlements: Jesus's uncle, Joseph of Arimathea ... Blake's hymn had referred to it with the line, 'and did those feet in ancient time walk upon England's mountains green'. The Brits liked their history. He was going to need all their help for, in his long life in US government service, he had often been confronted with strange sights ... but none as strange and incomprehensible as this one.

How was it possible for the remains of one of America's most secret aircraft to be lying half-covered at the bottom of this goddamn trench, in this godforsaken place, 1,500 years before it ever flew? Before anything ever flew? And, looking at the charring marks on the 'artefact', it had all the hallmarks of probably being shot down by a guided missile at a time when people only had bows and arrows, maybe not even those.

Right from the start he would have known at a glance what the object was, even if he had never been told. That was the easy bit. It was the front escape module of a sub-orbital aircraft containing its two-man crew, along with four passengers. Its overall 'stealth' shape would have been painted matt-black, something to do with radiating frictional heat as it passed at high speed through the air. The shape and paint also made it virtually undetectable by radar.

The module was designed to detach from the main body of the aircraft at a time of emergency, then parachute itself back to earth, even from the fringes of space. It was loaded with homing beacons and marker dyes to help searching ships and aircraft in its location, along with the rescue of those on board. He thought the company who had built her, probably Lockheed's famous Skunk Works, would be gratified to know their design had not only worked well but had also withstood what was estimated to be 1,500 years of environmental corrosion. Corrosion? It would never corrode. How could it? For to withstand the 2,000-degree external air temperature at which it operated it could not be built of metal, not even titanium, but instead had been manufactured from a plastic and ceramic composite which would never corrode in a million years, let alone a mere 1,500.

Coleman shuffled his position in the soft beach sand and squelchy mud to get a better view of the entry hatch. Three other military personnel had jumped into the trench and were now partially obscuring his view. Probably part of the general recovery team, he thought. They had moved over to the entry hatch and he could see that one of them was talking to one of the investigators still inside. Occasionally the one investigator doing the talking, and whom he could still see, would look in his direction.

Through all the time AURORA had lain undiscovered her structural integrity had been maintained, he thought. She'd been a good bird, he could see that. Sure, she was muddy and shitty; looked as if she had been badly scorched in a way that hadn't come simply from frictional heating. There was the recently opened escape hatch where the British had crudely forced an access. The faded white lettering above the hatch was still readable:

AURORA VI
High Altitude Escape Module.
United States Air Force
CXFM130508

His mind kept going round in circles, asking the same question and getting the same answer. He felt the utter strangeness, the unreality of the situation. He didn't believe in the impossible. There had to be a rational explanation. It was just that, at that moment, he couldn't see it.

He looked away from the immediate scene towards the car park, from where one of the young British military personnel was heading in his direction. Coleman glanced at the man's name tab. It said Captain Mark Lundis.

'Nothing inside?' asked Coleman.

'Not much, apart from what we have stored in the truck awaiting you or your men's collection. Nothing ... apart from the remains of a couple of bodies and, of all things, a cowboy hat.'

'A cowboy hat?' asked a surprised Coleman.

Lundis replied, 'Yup, obviously someone who enjoys the "wide, open ranges of cowboy country".'

'Any electronic equipment? Laptop or peripherals?'

'There's a battered laptop in good working order and some clothing. Nothing else. No peripherals.'

'I'll come back with you and check what you have. I take it your guys have bagged everything to prevent contamination through handling?'

'Of course.'

'Okay, I'll come back with you and leave these guys here to get on with their job.'

As they walked towards the car park and the Army command vehicle Coleman asked, 'What about the few remains?'

'What you'd expect of someone dead 1,500 years. There's not much left.'

'Was there a cause?'

'Bit early yet. Bearing in mind too they could belong to your government. We're not sure which remains are indigenous to the site – as in, Celtic in origin – and what, if any, might belong to the aircraft, like crew or passengers.'

'One or two bones look as if they might have been in a microwave. But don't take what I'm saying as gospel. It's early days and we have little to go on. All we can say for definite at the moment is height, weight and probable cause of death.'

'What about DNA testing?'

'Soon, according to our experts,' replied Lundis. 'A coupla days is my best guess.'

As they climbed the vehicle's few steps, Coleman's mobile began to buzz noisily. He halted and reached inside his overcoat pocket. Captain Lundis, climbing the steps ahead of him, also stopped. Coleman motioned him to continue as his conversation might be confidential.

After Lundis had gone ahead and closed the door of the command vehicle, Coleman answered with a perfunctory, 'Yes?'

It was Sir Freddie Stirling, speaking from the Atkinson Morley Hospital in Wimbledon, a London suburb.

'Those four casualties we saw at the Atkinson Morley the other day. You know they were all at Pendragon Point,' Sir Freddie explained. 'All are still unconscious although two of them, the boy and his nurse, were semi-conscious at first and then suffered a relapse. Pendragon Point is close to where you currently are, I understand. The family had been there on Christmas holiday for a couple of weeks.'

'But here's the interesting part,' Sir Freddie continued. 'According to one of our army captains, in an earlier conversation he had with the woman before Christmas she gave him precise information about what AURORA contained. She'd specifically mentioned two computer discs. The captain read from his notebook to me and I quote him: "There were two hard-drive external computer discs on board. We thought they were amulets worn around the neck. They were inscribed 'Property of the US Government'." She muttered something about the fifth century too. Now, I assume this will mean something to you. It clearly means nothing to us. They may be what you're looking for?

'Meanwhile, we have no idea when they might recover, but according to the specialist team with them, it could be quite soon. There's some new drugs they're going to try which look promising. Might be useful if we meet again, say in three days' time.'

Coleman's mobile went dead. He'd been given something to think about. It might not be much, but it was the only lead he had as to where the two Code Black discs might be. And how did she know about the two discs? He had to find out. Her information was too specific.

He resumed climbing into the command vehicle and opened its rear door. Once inside, the fluorescent lighting gave everything a stark appearance. Lundis drew Coleman's attention to a large table on which reposed a tagged, clear plastic bag enveloping a cowboy hat. It would have looked like a joke if it had not been so serious. It was enough for Coleman.

From the photographs he'd seen it could only belong to one man, field agent Tom Santos. He picked it up from the table, still in its tagged, protective plastic evidence bag. He turned it over and looked inside. 'John B Stetson', it said; an original. It fitted the man's profile; a profile he'd read and reread many times. He'd been on board all right. What further proof did anyone need?

He replaced the hat back on the table and remained silent, lost in too many questions, with no explanations for any of them. All he thought was that Tom would never have left the hat behind. Not voluntarily. Not without a fight. Then he thought of the human remains and wondered grimly whose they might be.

'What else do you know?' he enquired of Captain Lundis.

'Very little. We've not had enough time.' Then, sensing that this answer was not enough on its own, he added, 'I'm no expert, but it looks like a missile strike which brought it down. You can see from the burn marks.'

'How was that possible?' Coleman asked.

The soldier shrugged his shoulders. 'Search me. I don't think anyone has a clue.' His face remained impassive, before he continued: 'When this came down – or was brought down – the Celts hadn't even got bows and arrows, let alone surface-to-air guided missiles.'

Coleman picked up on the phrase 'guided missile'. Where did that come from? What had made the soldier venture that information? There could be many reasons, admittedly all pretty unlikely. Maybe a mid-air collision? What made the soldier think it was a missile? Coleman had a feeling he knew more than he was saying, but he decided to let it go and instead picked up on the word 'Celt'.

'Celts?' he asked.

'Yes. It's a Celtic settlement, probably belonging to an armourer, someone who forged swords. They were often mistaken for wizards since their knowledge of metals was highly prized, and their kings used to keep them apart, safe from capture by an enemy.'

Coleman didn't want to hear any more. He cut the soldier short as he asked: 'Is there any rational way this plane could have got here?'

'None. There is no way. Superficially, however crazy it sounds, it looks like it was shot down 1,500 years ago.'

Coleman caught that phrase: 'shot down'. Now he knew for sure that the soldier knew more than he was letting on.

'It couldn't have been,' retorted Coleman.

'We know that. But there is no other explanation.'

Coleman remembered the line from a Sherlock Holmes story: 'Eliminate the impossible, then whatever remains, however improbable, must be the truth.'

'The only thing I know for sure at the moment,' Captain Lundis said, 'is that we've told your team there's a possibility of flooding from high tides at this time of year. So tomorrow we've been asked by your recovery team if we can dig the vehicle out and, using a pretty big crane, lift it on to the quayside.'

∞

It was the end of the day and nearly everyone had gone from the quay and from the recovery site. The same helicopter that had brought Dexter Coleman to Pendragon Quay had now taken him back to London.

It would have been impossible to see if it had not been for the artificial lighting. Only four people were left on the quay and all were army personnel. Two were part of the overnight guard posted to protect the site; the two others were Captain Lundis and an army lieutenant.

'He had to know what we found, I suppose,' said the lieutenant.

'Oh, yes,' Captain Lundis replied. 'There was no way we could keep it secret. Not from them. Not that aircraft. He had to know. It was orders that we had to keep nothing back, apart from the wash-down tomorrow.'

'Wash-down?'

'Yes. Should be easy. Routine. Within forty-eight hours the Americans should have their own crash recovery team here. Must say, overall, they seem to have been a bit slow with this one. As if it's not a priority for some reason. Can't quite put my finger on it.'

'Forty-eight hours!' the lieutenant replied. 'That should be more than enough to scrub the evidence clean, to get rid of any explosive residue there might still be. It's amazing. We've done the forensic work and after all this time it's still there ... after more than a thousand years.'

'After more than a thousand years,' echoed Captain Lundis. 'Amazing for an RDX-derived explosive like the PBXN-109 used in missiles. These days most military explosives can be traced to their source. They're chemically fingerprinted so they know the batch, date made, who ordered it and all the rest. We've done the forensics and yet it's still possible to trace this batch of PBXN-109 to the Ministry of Defence, the Royal Navy, and then to HMS Daring. So tomorrow we scrub whatever evidence there is, as clean as a new pin, with high-pressure steam cleaning.'

∞

Consultant Neurologist Dr Hugh Orum and neuroscience professor Richard King from London's Imperial College were relaxing in their chairs for the first time in two hours. Their two visitors had just left the Atkinson Morley's new Nuclear

Imaging Department. Orum had just returned after following his visitors out in order to have a quick cigarette in the open air. One of the visitors was a slightly mysterious government official called Sir Freddie Stirling, together with an equally mysterious American named Dexter Coleman. Orum and King had both noted that neither had given their professional capacity or why they were making their visit. However, both had been given credibility by the fact they had been accompanied by the hospital's medical director, who had behaved even more unctuously than usual.

The director introduced Sir Freddie as having 'a special interest' in the patients lying unconscious in the adjoining small ward but declined to give details as to what that interest was. Meanwhile Sir Freddie's American guest had been introduced as merely a 'Mr Coleman', with no further details volunteered. They had both asked a lot of questions, mostly about when the two doctors expected their patients to revive. Because of the Atkinson Morley's role as a 'teaching hospital', visitors asking questions were the norm rather than the exception. But added zest was given with the appearance of two 'security officials' who stood silently at the end of the corridor outside and were on guard round the clock. The medical director explained they were going to be a permanent feature for the foreseeable future.

One of the two computer screens Orum and King had just been watching had gone blank, while the other showed what had been, until a few moments before, a continuously moving number of separate graph traces, depicting various channels of brain activity.

They looked at each other in mute amazement. King was first to break the silence. 'Now we've got rid of our guests I think we can be quite frank with one another about what we've just been seeing,' said Orum.

'Well, what can one say? replied King. 'We were told by Los Alamos that the results might be good. They didn't say 'spectacular.'

'Quite breathtaking,' Orum replied.

King leaned back in his chair. 'That was Ben?'

'Yes,' said Orum, still reflecting on what he'd been watching. He added, 'The amazing thing is that what we've just been watching was recorded off-line on to several DVDs. Since then the equipment has been improved so we can watch in real time … as it's actually happening, as it were.'

They both fell silent for a few seconds, each lost in their own thoughts. King broke the silence. 'The pictures of what Ben was actually seeing, his mental pictures, were stunning; moving pictures, too. The many flashes and seeming interference is interesting, too, particularly when correlated with the readings on the other screen. It's like watching a faulty television, and yet is actually what is going on in his head.'

'Ironic,' said Orum, 'when here we are watching these, that yesterday he was back here with his family, air-lifted in by the army. Some accident they were involved with while on holiday, we were told. Oddly, when they arrived all were unconscious except for Ben and his nurse. They lapsed into unconsciousness a short time later. Dunno why they were different. Either way, here they all are. The army chap who accompanied them seemed interested that Ben was already a patient.'

'We've checked them all out, of course. Brain-stem activity, as far as we can tell, is normal. Indeed, there's a great deal of subconscious brain activity going on. Plainly they aren't in a persistent vegetative state, but for some reason aren't conscious either. It's as if they're all in a coma that they could come out of at any moment. How they got this way, and why, we're evidently not allowed to ask. I think there's a lot we're not being told. Two security folk on the ward door, and odd people coming and going. Some of those look "security" too. You can tell from the shape of their shoulders and their clothing. The Sir Freddie chap whose just gone is a minor cabinet minister. He keeps coming and going with the rest of them. He was on television news a few nights ago.'

'I think we can assume, Hugh,' said King, 'that it's something to do with queen and country.'

Orum replied, 'I got talking to one of the guys who came off the helicopter that brought them here. I was part of our usual medical greetings team. I recognised Ben immediately. Then there were his parents, plus one of his carers.'

'It was shortly after, when this politico arrived, this Sir Freddie something. I told him I knew the family.'

Orum leaned forward and picked up s sheaf of case notes from the desk in front of him. 'Ben's been a patient of ours for some time. The parents believe, and the hospital concerned has accepted, that his condition resulted from medical negligence when he was born. It was only a week or so ago that we decided to use him for our new imaging facility.'

'From what we've just seen, he's very special in giving us these breathtaking images …'

' … But where do they take us?' interjected King.

'That's the toughie,' said Orum.

CHAPTER 23

It was just after noon the following day, January 3rd. 'The American Embassy, Grosvenor Square?' the chauffeur repeated in a low voice.

'That's it,' Coleman replied perfunctorily. The black Jaguar saloon care of the British Government's pool car fleet, pulled away from the pavement outside the main entrance of MI5's headquarters at Thames House, in London's Millbank district. On the opposite bank of the River Thames stood the brown stone flanks of the Houses of Parliament.

The early morning meeting with Sir Freddie Stirling had been how he liked them: short, sweet and to the point. They had discussed what Coleman had discovered while at the 'crash site', and Sir Freddie had provided more welcome news about 'the patients', who were currently still unconscious at the Atkinson Morley. Sir Freddie, relying on the hospital's educated guess, was hopeful that they would make a recovery 'any time soon, certainly within the next couple of days'. It was then a question of asking how 'the woman' knew about both the aircraft and the missing discs. They had arranged to meet later that afternoon, at the hospital.

'Your driver will know how to get there,' Sir Freddie had said. Coleman knew it was all a long shot, but still better than nothing.

It had been a while since Coleman had last been in London and he was glad it had changed so little over the intervening years. As a result he believed he could still find his way around. But last time, he remembered, he'd been a 'foot soldier' without the benefit of a chauffeur-driven car. Now it came with the territory.

He had been particularly pleased with the 'luxuries' his status allowed, particularly the previous evening, when 'his'

Government-chauffeured car had been waiting. He'd been tired from that day's early start and then the long day, mainly waiting around for other people to do something. And he hadn't catnapped on the flight back. It was when the helicopter had touched down at the Artillery Club's cricket ground, close to London's financial centre, that he had been thankful for the things that came with executive power.

Despite his instructions to maintain daily contact with the White House, the pressure and varied activities involved in his current task had made it well nigh impossible. When vital, he'd maintained contact with Wurzburg using his mobile. But the non- secure nature of it meant conversations were guarded and always kept to a minimum.

Coleman reviewed the messages on his mobile. Most were from Wurzburg, most full of profanities and demanding to know where he was, and why he hadn't been in touch. An early morning start had again prevented use of the secure room at the American embassy before leaving. This had merely added to Wurzburg's ill-humour. Coleman knew that in these 'operations' communication was essential, otherwise seven tonnes of shit were likely to fall. And he knew who the fall guy would be.

It meant a conversation with Wurzburg had risen to the top of his list, even ahead of his major priority to find the discs. And there was precious little time left to do that, however he played it.

The US embassy occupied the whole western side of Grosvenor Square, with its roof-mounted stone eagle casting a beady eye on all who passed below. Usually, there were large numbers of people milling around its front entrance and wire crush barriers, or quietly queuing for a visa or passport. It was usual too for protesters to be agitating about something or other while holding placards proclaiming their message about suspected US involvement, or lack of it, somewhere in the world.

Although it was a nine-storey building only six were visible, with the other three below ground. A city ordnance made at the time of its construction in 1960 had ruled its overall height could not exceed that of other buildings around the square. The sixth floor and top floor housed the ambassador's penthouse, while the below-ground second floor was where the embassy's secure room lay.

Coleman phoned ahead to let the embassy know his estimated time of arrival at 1.15 p.m., then replied to one of the many texts from Wurzburg, to let him expect an imminent communication.

Coleman stopped momentarily to show his pass to the two embassy security guards before taking an elevator down to the second floor. He thanked God it had been redeveloped since the last time he'd used it. Now it was more starkly furnished than ever, save for one chair and a table on which stood a computer screen and keyboard. It was bizarrely simplistic for a time that had a high-tech solution for everything. He touched the keyboard, the screen came alive, and he typed in the coded address. Within seconds he was through and the image of Cliff Wurzburg was before him. He looked tired; like someone who had not recently had much in the way of a good night's sleep. But his response suggested he was still at the top of his game.

'Hi, Dexter. What you got for me? Time is something we've not got much of. So what's going down? Make it good. The president needs good news. But not so badly you've gotta make it up. So let's not fool ourselves,' he warned.

Coleman remained calm, guessing the stress Wurzburg was under. 'I might have a lead,' he said, 'but won't know for sure for three days, until the 6th. But it looks good at this stage.'

'That's what we wanna hear,' said Wurzburg in a deadpan voice. 'So what's it about? Make it good. I gotta tell the president something positive. Otherwise, soon, real soon, he's gotta be doing something. Like calling in the Seventh Cavalry. We're outta anything else.'

Dexter tried to put as hopeful a face on it as he could. The woman he was to see in three days' time just had to know something, he said. 'The information she had just couldn't be guesswork. The discs and the aircraft? It's too precise and there was too much of it. So how can she know what she knows?' Coleman paused to let it sink in before continuing with, 'That's the good news. The bad is she's still unconscious at this time. The medics are working on them all as we speak. They think in three days it will be as good as it gets. Otherwise we could be in for the long haul, and outta time.'

'So what's next?' asked Wurzburg. 'What do I tell the president?'

'Tell him we're hopeful,' responded Coleman. 'Tell him more than that, tell him we're very hopeful and to lay off calling in the Seventh Cavalry.'

'He said he was giving me until the fourteenth. It's only the third today. Another three days takes me to the sixth. He's gotta give me that time, Cliff!'

'Okay!' responded Wurzburg. 'I'll work on him. Meanwhile, there's things I gotta tell you from this end.

'First, is that the US Geological Group's Special Working Party has confirmed we're gonna have earthquake trouble in Washington state soon, real soon. They've detected deep cracks opening up already around Hanford. By deep I mean like twenty-five miles underground. Too deep so far to impact on what Energy is doing with their clean-up plan of the whole Hanford site, or on our own DEEP EARTH project. But it's gonna happen. It's keeping the president awake, I can tell ya.

'The slightly good news, if we can call it that, seems to be our Iranian expert. He's still right on the button with regard to his latest predictions, so far as we can tell. So he was doing a good job for us right up until your guy Santos found him out at Alamogordo. Pity he's now languishing in one of our jails in New Mexico. We assume Santos asked the local police to hold him there until he could be questioned when he got back from the UK. Clearly your guy Santos had no idea what he was getting into.

'This throws up the whole question of what was Pahlavi doing with the two Iranians? Could be the usual blackmail stuff. But did they force him to tamper with what he was doing for us? Give us false data for DEEP EARTH? We don't know, and won't know until you discover the discs, if ever. So what you need to do is find 'em, then question Pahlavi further to check whether the data is still kosher. We don't know at this stage whether Pahlavi will be a wimp, or a toughie to crack.

'But however we look at it we still need the stuff you're lookin' for in order for any of it to be worth more than a row of beans. That's still gotta be first base for us. Without those discs we can't crank up the whole DEEP EARTH shooting match, because those missing computer discs should tell us the wheres and wherefores of drilling the tunnels we need to have in place for our DEEP EARTH nuclear bang.

'Those tunnels will connect at one end with the huge cavern our underground nuclear explosion will create and, at the other, with all the toxic shit we've got stored near the surface. Then, according to our own geologists, the earth's natural subduction movements will take it all away. Eventually down into the earth's molten core. It'll take tens of thousands of years, of course. But we're in no rush. While the Earth's doing its job of recycling, it'll be safe. That's the main thing. And thank God we'll have the environmentalists off our backs who want us to get rid of nuclear waste, but not if it's in their backyard.

'Let me say again, without those discs we're in the dark, and would more than likely make an already big mess a whole lot worse. I won't go into any of the technical stuff. I don't understand it and probably neither would you. But it won't change anything I've just told you.

'Of course, we've got contingencies. None of them are going to be pleasant. The president is between a rock and a hard place. Soon, we're gonna have to press the button on these back-ups unless you can pull an ace outta the hole. We need it.'

Wurzburg made a movement to turn the screen off as if the conversation was over. He stopped himself.

'We've tried to get the Third Air Force to move slowly on AURORA so they won't get under your feet with what you're doing for us. According to the British, it's been stuck in the Cornish mud for over a thousand years, so a few more days ain't gonna hurt. But you know the military, Dex. There's rules covering every goddamn situation. We can only stall 'em for so long.' He paused and seemed to be turning the matter over in his mind before continuing. 'Yeah, AURORA's a mystery all right, but she's not currently our – or your – major priority. You do your job and let the military do theirs.

'Which reminds me,' he continued. 'There's one other thing.' He paused, then said: 'This woman you've talked about. What's her name? Anderson? Katrina Anderson? And who maybe has the answers to some of our questions?

'Can you give me a few more details? Maybe I can get our immigration authorities to check her out? We get close to thirty million visitors a year into the US, excluding the Canadians and Mexicans. She mighta been here and show up through having answered our ESTA immigration visa requirements. Since 9/11 and their Brit "shoe bomber" we've got a lot tougher on handing those things out to our British friends. I can get our guys to check it out and maybe give you a heads-up, ready for your meeting, when was it for? The sixth, was it?

'Yeah, in three days',' said Coleman in response. 'That's still the latest I've got.

'It's a long shot,' continued Wurzburg. 'But "long shots" seem to be all we've got to work with these days.'

Dexter agreed, and gave Wurzburg the information he needed on Katrina Anderson. Yeah, he agreed, 'long shots' did sometimes come off.

The computer terminal's screen went blank and he was left with his own thoughts. Not much had changed, Wurzburg's information included. He now had 'the Anderson lead' though, care of the Brits. He might need back-up from them at some point, so he knew he had to tread carefully. In the meantime, he turned the name of Katrina Anderson over and over in his mind.

∞

'Hello, Jack,' Sir Freddie said as he entered Sir Jack's office. It was late afternoon and growing dark. Sir Jack's back was to him and he didn't respond immediately or turn round to acknowledge his presence. Instead he continued looking out from his office window, out along the old runway to where yet another articulated truck had been turning so that its rear roller shutter doors were adjacent to the roll-up door of the hangar. He remained silent for a few seconds more as he watched the hangar door rise up slowly to reveal an opening thirty feet high and forty feet across. A forklift truck appeared from within, ready to unload heavy equipment off the truck.

Sir Freddie Stirling ignored Sir Jack's calculated rudeness by ploughing straight on. He too, decided to ignore politeness. 'You know why I'm here, Jack, and it's not for a debating match. I'm going to tell you what to do and you're now in the business of doing it. No questions.'

'And if I refuse?' Sir Jack replied over his shoulder.

'Then you could be leaving. There are two security guards stationed outside your office door. I could ask them to escort you off the premises. They would do it right now, not tomorrow or the day after, but right now. Your deputy would take over.' He paused to let his words have their effect before continuing. 'But I'm hoping good sense will prevail and we won't need to use that sort of unpleasantness.'

'What do you want me to do?' asked a mildly curious Sir Jack in a monotone.

'Actually, I don't want you to do anything. We have a plan for containing AURORA, and I think you'll like it.'

Sir Jack was no novice in the political gamesmanship arena usually involving large engineering projects such as BRIGHTSTAR. It was part of their appeal to him. But he remained a realist. He always remained that way, come what may. He always knew where his authority started and ended. On the one side was the running of the project. That's what he did. On the other side was Government politics. That was where the real power lay, where the real project bosses lurked.

He knew that however one cut the cake, he was the man in day-to-day charge of BRIGHTSTAR; the captain of the ship. When something nasty happened it was part of his job to manage it, part of the risk he was paid to accept.

But beyond that were his political masters, the ones who footed the bill – and paid Sir Freddie to do their bidding. He, Sir Jack, might be the man in charge, but he never lost sight of the simple fact that they were in charge of him. That was the downside. The upside was it was just a job, not life or death. At the end of the day he worked to live, not lived to work.

'So what does Carr's preliminary report say?' asked Sir Freddie, adding: 'It can't be much other than based on educated guesswork. Not been enough time for anything else.' He flung his coat over the back of a nearby chair close to Sir Jack's desk and sat down, still facing Sir Jack's back.

'I'd already left this morning,' Sir Freddie continued unabashed, 'before my own copy arrived. But I knew you'd have one I could crib off.'

Ignoring the 'No Smoking' sign on the desk and situated next to an ornate, glass ashtray, Sir Freddie pulled out a large Havana cigar and said, 'Hope you don't mind' as he went ahead and lit it.

As he blew out a stream of dense smoke he said, 'Of course, there's been whispers, so I already have more than a clue as to what's inside it.'

Sir Jack at last turned to face his interlocutor. He could have walked to his desk, retrieved his own copy of the report and tossed it, perhaps theatrically, over to Sir Freddie. He could however almost recite its contents off by heart, for Sir Jack had a photographic memory. He judged, too, he didn't need theatrics to score points at this stage of the game.

'Before we go there, Freddie, what do you want me to do?'

'Nothing, I've already told you.' Sir Freddie looked comfortable and inhaled more cigar smoke. 'Do nothing, Jack. Keep your head down. Let the bullets fly – or in this case the rockets – and land where they may. Now tell me about the report.'

'It's called the "Preliminary Report into the Daring Incident", authored by Professor Sir Robert Carr, special advisor to the BRIGHTSTAR project. Not a very snappy title, but I think it'll do,' Sir Jack said in a low voice, devoid of humour.

'He wraps it all up in a load of science and maths only a few people in the world could understand. But basically the message is clear. An earth tremor might have set the whole accident in motion, but the fact remains that our friend, Dr

Jonathan Anderson, was probably right all along. Carr concedes that much. We never understood what effects BRIGHTSTAR might have on space and time. It's all down to Einstein.'

'He now says, too, that BRIGHTSTAR was not only at the edge of physics as we understand it, but may well have been beyond its leading edge. Now he's had time to look afresh, he says it was always inherently unstable; the beam was always capable of drawing its own timeline, independent of the real world timeline it was actually operating in. In other words, in certain circumstances, it could operate like a time machine at a quantum level.

'He goes on to say this effect could be dangerous for anyone caught in the beam. He cited two potential areas of risk as the beam swayed back and forth, due to the effects of the earth tremor.

'One was across the bay at Tintagel, the island part that juts out into the sea and which they try to get everyone to believe was once home of King Arthur.'

Sir Jack was staring straight into Sir Freddie's eyes, trying to gauge his response. 'The other is at Pendragon Point,' he continued, 'where the old lighthouse is, and where the Anderson family appear to have been on Christmas holiday. The beam might have skimmed its upper part. The operative word here is "might". We don't know for certain.

'Then there's HMS Daring herself. She was part of the experiment to find out how one of our most modern warships and its crew would cope if they were illuminated by our beam. In theory we expected little to happen. The point of the test was we were trying to extend BRIGHSTAR's range beyond the horizon using a mirror system suspended from one of our helicopters. We bounced the beam off the helicopter flying at a carefully predicted altitude and distance and from there onto Daring. Simple optics really. In theory we were merely following something we'd already done with our established capability linking Morwenstow with Glastonbury.' Sir Jack could see he had lost Sir Freddie with his explanation.

'You *did* know about the countrywide plan for BRIGHTSTAR?' Sir Jack asked.

'No. Can't say I did.' Sir Freddie replied before adding. 'Remember, I'm the new boy around here.'

Sir Jack continued almost without acknowledging Sir Freddie's response. 'BRIGHTSTAR was meant to provide protection from rogue nation ICBM rocket attack for the whole nation. This was going to be done through a network of optical repeater stations dotted around the country. The first is already established in the Menwith Hills close to Glastonbury. It's worked well during earlier tests although it wasn't operational during the night in question.

'It seems to have been Daring where the beam's effects were felt most and possibly resulting from the earth tremor. All we know for certain, Carr writes in his report, is that it caused Daring to lose a missile, either by affecting the men, the electronics, or both. We've got to assume it was that missile that brought down the American aircraft.'

'Ah! Yes. The American aircraft,' Sir Freddie echoed through a cloud of cigar smoke. He continued, 'we've got one of their guys flat footing around here as we speak. Appears they've lost something important. But won't say exactly what apart from the fact it appears to be two computer disks.'

Sir Jack added, 'Carr says the American aircraft was bad luck. As our beam momentarily wavered around the sky it flew right into it. It was in the wrong place and at the wrong time. After that it's, well, anyone's guess.'

Sir Freddie had been looking pensive before he said. 'But whatever happens, we don't want suspicious Americans pointing their fingers in our direction. So we're giving them all the assistance they need short of being actually helpful.' Sir

Freddie's grey eyes then stared unflinchingly at the broad figure of Sir Jack as Sir Jack walked from the window and over to sit behind his desk. Sir Freddie then inquired, 'Tell me, Jack, did Carr mention the "Bell Island mystery" anywhere?

'Nope. That's a new one on me,' was Sir Jack's immediate response

'I'll come back to it shortly,' replied Sir Freddie. 'But first let me ask you another question. 'What does Carr say about the earth tremor?'

'He says it's unlikely we could have done anything about it. It was one of those unpredictable things; an act of God. It didn't much matter where GCHQ Bude was sited. The fact is that although the UK is relatively non-seismically active, as geologists say, that's still a long way from saying it's tectonically inactive. There's nowhere on earth one can say is positively safe. Period.

'The earth's crust moves constantly. For the UK, the effects are usually twenty-five miles below the surface. Sometimes ... rarely ... we feel it especially in Cornwall for some reason. There's no way of predicting when it might happen, or what its effects might be. It was our bad luck for it to happen on the very first night we ran BRIGHTSTAR up to full power.

'According to Carr, it defocused the optics and sent the projector out of alignment. A small movement at our end, of thousandths of a millimetre, sent the beam oscillating wildly by maybe a few feet at the target end; bit like waving a garden hosepipe around.' He paused to look into Sir Freddie's impassive face.

Sir Jack added defensively, 'Besides, Morwenstow's siting had been long fixed before I arrived on the scene, probably before I was born.'

Sir Freddie knew what he was hinting at. He laughed a hollow laugh. 'If there's a witch-hunt, Jack, which I don't think there will be, I don't think that argument would save you. You were captain of the ship at the time it appears to have struck a rock. That's the plain, simple fact of the matter. So you'd be the guy with the noose round his neck.' He paused a few moment for effect before adding, 'Besides it's in part what we pay you so well for.'

Sir Jack knew he was right. If the Government wanted, he could be toast at a time of their choosing.

'But let's not dwell too long on the negative aspects of this,' Sir Freddie said. 'Shooting down friendly aircraft is not part of our defence policy. The Americans do it all the time. Ask our boys in the army. They've often been at the wrong end of American gunnery practice.'

'Funnily enough, the Americans seem to be playing a slow game on this one. We expected half the American Third Air Force down about our ears by now. But no. They're far more concerned about the computer equipment the aircraft had on board when it left but which doesn't seem to be on board now. We certainly don't have it, as I've told our American gumshoe. Coleman's his name. 'Why would we?'

'We've told the Americans we have heavy lifting equipment we can move in to recover their precious aircraft, and it had better be done sooner rather than later before a seasonal spring high tide either washes it all out to sea at worst, or at the very least fills it up with seawater. We've told 'our cousins' that if the weather gets very bad, which it can down that way, it could all get washed out to sea never to be seen again. And not to blame us if that happens. They were warned.'

'We've pointed out that Cornish weather waits for no man. The US no longer has the heavy equipment in the UK they used to. They've had their defence cuts too.' He looked up at the ceiling in mock exasperation. 'Pity about the ending of the Cold War. We all knew where we were then. Now, nobody knows anything except the enemy seems to come from the

Middle East where the name of the game is something called "asychronous warfare".'

'But I digress. The US defence cuts mean they would have to bring heavy equipment over from Ramstein. Getting it to say RAF Fairford would be the easy bit. From there it could get tricky, given the fact our narrow and high-hedged Cornish roads aren't best equipped for moving heavy equipment along. Except very slowly. And you know what it's like at Pendragon Quay? The road down to the quay is pretty aggressive even for a car. And that's without factoring in the hairpin bends. A sea operation too, would be equally difficult because of the rocks. A helicopter heavy lift operation too is out of the question because of the proximity of the cliffs. So, in my opinion, that leaves the Americans with just 'us' as their only option.

'Although I need to check this, I understand we've just been given the green light,' Sir Jack looked a little smug, 'to go ahead with our offer of help. This was after our friend Coleman went to Pendragon Point to see what it's like for himself. He's seen how the land lies together with his chums from the Third Air Force Accident Investigation Branch.'

'Coincidentally it gives us an opportunity to wash the aircraft free of any muck there might be clinging to the wreckage. Explosive residue and the like, which the suspicious might trace back to Her Majesty's Government.

'Once that's done,' – he looked at his wristwatch – 'which will be within the next hour or so, it only leaves us with one loose end to take care of. And that might involve us helping Coleman to find his precious lost discs. And that's it. I'm seeing Coleman again on the 6th,' – he looked again at his watch to check – 'that's in three days' time. It'll be when I'll try and sort out the lose ends with him.' Sir Freddie reached out and extinguished his half smoked cigar in the glass ashtray on Sir Jack's desk.

'Somehow, Jack, those two discs are connected with the mercifully few casualties we have in all this. There are only four, if I can call them that, and they all remain unconscious at the Atkinson Morley hospital. Interestingly, they all belong to your Doctor Anderson and his family. Now what was he doing down there? I'm too old to think it was mere coincidence.'

'Jonathan Anderson?' repeated an unusually surprised Sir Jack.

'The same,' Sir Freddie replied evenly.

'First I've heard of it,' responded Sir Jack. 'God knows.

I've no idea. Really. But before we can ask him anything, he needs to be conscious.'

'Your guess is as good as mine as to when that might be,' Sir Freddie responded.

'You mentioned Bell Island …'

'Ah, yes,' said Sir Freddie, remembering. 'The 1978 Bell Island mystery. It was in a BRIGHTSTAR briefing document I was given when I flew down there for the, I think penultimate test. I assumed Robert had written it.'

Sir Jack replied: 'He did have a large input, but I can't say whether he was responsible for that bit or not.'

Sir Freddie continued, 'It's well documented, but still wholly unexplained. The Canadian Bell Island, out near Newfoundland, was where a huge boom was heard by people living up to forty-five kilometres away, and it caused a fair amount of damage. Conspiracy theorists had a field day. Still do. US satellites picked up a flash, which they said was brighter than the Hiroshima atomic bomb. That was about 20 kilotonnes of TNT if I remember correctly. Some tried to link it with ball lightning, despite weather conditions not being suitable for it. Others linked it to a suspected US Defense Department's directed-beam energy weapon. Interestingly, Brookhaven National Laboratory, on Long Island close to

New York, isn't that far away. Maybe it was linked, maybe it wasn't.'

'So what's your point, Freddie?'

'My point is this. Bell Island may or may not have been the Americans messing around with a directed-beam weapon, similar to BRIGHTSTAR.' He paused to let the information sink in.

'So, I'm suggesting we wrap up the loss of their AURORA in our own mystery ... just like Bell Island. The Americans might suspect more, but what are they going to do about it? Send in the Sixth Fleet? I don't think so. At the end of the day, they can take it or leave it. Fortunately for us, not all things can rationally be explained.'

'We say it's a mystery, and they'll have to accept it; have to accept that not all mysteries can be explained. That's why they're mysteries.

'I think we'll get away with it, but I don't want you upsetting things by doing your own clearing up. We're at a delicate stage. Another couple of days, maybe less, and we'll be in the clear.

'The only slight concern I have is with this chap Dexter Coleman. He's head of their NID. Now why would they send him to investigate the loss of a couple of computer discs? It's more CIA who deal with overseas matters. NID is an internal US operation. Something doesn't square here. I don't think the Americans are playing with a straight bat, Jack, when it comes to their precious AURORA.'

∞

Dexter Coleman typed in the security codes needed before the screen lit up. Earlier messages had come through on his mobile saying Wurzburg wanted to talk and talk urgently, but it had to be via the US embassy's secure room.

'Hi, Dex,' the face of Cliff Wurzburg said. 'Got some stuff I thought I'd better get across to you soonest. It concerns your

Katrina Anderson. Remember our discussion about her?' He paused before asking: 'I take it it's all still on with her?'

'Sure. I've been told nothing by the Brits to make me think otherwise.'

'Good,' said Wurzburg. 'I'll save the best till last.'

'Oh?'

'First is, our Immigration Department have come up trumps on Mrs Katrina Anderson. She's quite a star when it comes to ancient history. Been to quite a few symposiums in the US. Some stuff at Harvard, Washington and at Caltech. Her husband too, a Dr Jonathan Anderson. He seems to have been quite something, but in recent years he's faded from the "theoretical physics scene". But get this … their son, six-year-old Ben Anderson …'

'How can a six year old help us?' interrupted a surprised Coleman.

'…he's part of a nuclear medicine outreach programme being run by Los Alamos. Yup, the same Los Alamos who own the two lost discs we're looking for.'

'How does that help?'

'Because less than a month ago he was hooked up to a new medical imaging facility. The Brits are helping to pioneer it with Los Alamos. It's all to do with an emerging medical imaging area that can help patients with severe mental health problems.

'These guys are based at somewhere called the Atkinson Morley hospital. Evidently, they use this technique to sort out mental problems and stuff with the brain, and can peer right inside your mind and get images out. They've got pretty good at it. And ain't you going over to the Atkinson Morley in a coupla days? Isn't that where the Andersons are now? Unconscious?'

'Those are the ones you're hoping to get some answers from? Don't know if God suddenly loves us, but maybe you should be asking your new British pals if they should apply this technique right now with the whole Anderson family? The emphasis is on the word "now". I know it mightn't sound ethical. You know what doctors are like when it comes to ethics. But, son, we don't have a lot of time to worry about ethics right now. We've only got nine days before the president has to start up other options.'

'It's why I'm sending our Dr Juan Ortez over to help you out. He's the leading guy from the Los Alamos nuclear medical imaging programme. As we speak, he's on a flight heading in your direction. He'll be with you by 8.30 a.m. the morning of the seventh. Earliest he can do it. At present he's outta the country. At some symposium in Hong Kong. We've tried pulling all sorts of strings but it's still the seventh.

'Meet him at the airport. Look after him. He knows that what we're asking him to do is important, but he doesn't know why. Don't tell him. He knows all he needs to know, but what he doesn't know about his speciality subject ain't worth knowing, they tell me. So use him. You got that? He already knows we wanna "jump start" your patients.'

'But we don't wanna frighten the horses so the Brits and their medicos get themselves a head of steam. What they're doing right now could help get us out of the DEEP EARTH logjam. We need to keep 'em on side.'

'Something I need to tell you but is for your ears only. The military have used him already with some of our detainees from Al Qaeda. He gets good results the easy way. Waterboarding and Guantanamo could be old hat now.'

'I know what I'd do if I were you son. But you're the guy on the spot. Must be your decision. Don't let me pressurise you into anything. But while you're thinking things over, think too of the twelve million Americans whose lives you swore to protect when you got your badge of office.'

The screen went blank.

CHAPTER 24

It was early morning. Dr Orum and his young medical assistant were near to completing their routine round of bedside tests on their mystery and still unconscious four patients. It was just as they had been doing on a daily basis since their arrival nine days before aboard a military helicopter.

'I don't like it,' Orum's assistant was saying. 'It's okay for our medical ethics committee to say we can go ahead. But it's not their head on the chopping block if it all goes wrong.'

Dr Hugh Orum looked unconcerned. His career had been long and he was therefore more experienced than his houseman. 'I take your point,' he replied. 'But your concern about the hospital not getting patient or next of kin permission will be the least of our problems, you mark my words.'

'Dr Ortez I'm told is an authority on both nuclear imaging and the unconscious mind,' Orum continued. 'At the very least he's a neuroscientist amongst his many qualifications. Our patients will be in very safe hands. As long as we're satisfied that we've obeyed our professional oath of "first doing no harm" then we'll be in the clear.'

Orum had just finished checking Katrina's pulse and was going on to do the same with Jonathan Anderson. He looked at his assistant and with a chuckle added, 'Don't worry. There'll be more than enough people wanting to batter the walls down over the whole question of imaging ethics per se. If any of this comes out there'll be those saying doctors are prying into people's innermost thoughts - which of course we are. Then'll come accusations of "The State" using it in place of torture to extract anything it wants to know.' Orum gave his assistant a gentle nudge and with a wink added. 'Look on the bright side son. Physical torture, thumbscrews and hot irons, will become things of the past. On the negative side all we've got against us is that our unconscious patients can't give us their permissions that it's OK. At the end of the day we're just doctors doing our best in using the latest technology to help those same patients. So my advice is we stick to doing what we're doing. Let those higher up in the hospital's command chain take the wrap, if there's gonna be one to take. Which I doubt.'

It soon became apparent to both of them that there'd been no change since the previous day. Their patients were still unconscious. All the repeated medical tests had done was indicate they were perfectly normal. Brain-stem activity was normal, respiratory functions were normal, autonomic functions were normal. As Orum had gone from bed to bed all he could add to each patient's notes were the letters 'GOK': doctor's shorthand for 'God Only Knows'.

Finally Orum and his assistant were finished and moved across the corridor into the Imaging Room. They sat down at a desk supporting a large screen linked to a computer squatting on the floor. This was linked in its turn to a high speed datalink service connecting directly with a partition of the Los Alamos Advanced Simulation Computer (ASC) facility sited 2000 miles away in New Mexico.

Orum was a whole lot less certain about what they were doing than he sounded. He would have liked to have talked further to his assistant about the ethics issue but had then thought better of it. An inner voice had warned him that he could have been treading on dangerous ground. Only a few days before, he and his team had been forced into signing a government 'confidentiality agreement' where they had been told 'no, its not a secrecy agreement, not the same thing at all.' One person had refused. She'd been moved away 'temporarily' to 'other duties' within the hour, 'on the orders of Sir Freddie', Orum had later learned.

Another warning factor had been the hospital's medical director and his high state of nerves. He was not the type normally. He left them to get on with their jobs whilst he got on with his. Rarely did he interfere. But the previous evening he'd been over asking them how these particular patients were. Orum had gone over the same ground he and his team had already been over more than once already. Narcolepsy had looked likely but was quickly ruled out, as was Kleine-Levin syndrome. 'There's fewer than a thousand cases in the whole world,' Orum had repeated, 'so it's unlikely. And not with four of them getting it at the same time.' He'd added too the same statement he'd made the last time when asked: 'It seems to us there's something we're not being told.'

The medical director had let slip that there were 'important political people,' taking a deep interest in their cases. Orum had mentally scoffed when the medical director had then sworn him to silence for his 'slight slip of confidence'. Any fool could have already deduced as much for himself.

For a start the patients had arrived aboard a military and not a civilian helicopter and yet were not military personnel. Orum had learned later that they'd been transferred from a Cornish Air Ambulance Service Eurocopter at RAF Fairford. Then they'd then been escorted into the hospital by medical personnel normally seconded to the military for exercises overseas. Finally there were still two military guards at the end of the corridor. 'It didn't take a genius to work it out' Orum had thought.

No one had said 'how' his patients had come to be the way they were, other than that 'they'd been involved in a military training accident involving a radiation leak.' Orum knew this to be bullshit. But why wouldn't anyone level with him? What was the 'state secret' everyone seemed so anxious to protect?

And who was the mystery American whom he'd only briefly met? What authority did he carry over anything? Finally, who was this Dr Ortez the hospital was expecting to arrive anytime soon?

The medical director had said he was a medical doctor and head of 'the imaging team' at Los Alamos National Labs. Clearly he was important. But why was he visiting them now? And who, or what, was Sir Freddie Stirling? Whoever he was, he seemed scary and clearly carried great authority.

Sitting at his desk Orum stared past his assistant in the adjacent chair and tried again to make sense of it all.

∞

Suddenly Ar-tur knew the answer. He did not care for himself, only for her, for Ganhumara. But time was running out. He screamed upwards into the empty heavens. 'Not now!' he pleaded, but his voice was lost among the sighing of the leafless branches and the gust of the chill, winter's air. 'Not now!' he shouted again. 'We need more time. Just a little ...' – his voice faded away to nothing.

It echoed emptily around the narrow track that had led him and his three accompanying soldiers from Camlan before they had lost touch with each other in the snow storm. His voice echoed dully out among the high oak trees whose gaunt and silvered branches looked like marching ghosts. It bounced off the frost-covered fields rolling gently away on either side of the ancient track that cut through them. It was out of this he then heard another voice, separate from his own yet somehow belonging to him. He knew it. Was it an angel's? Or was it a child's?

'Star light, star bright'

The voice began. Followed by:

'First star I see tonight;
I wish I may, I wish I might,

Have the wish I wish tonight.'

It was so familiar. Then it was gone. His body jerked with a spasm he assumed resulted from wounds he'd sustained in battle. His body convulsed before pitching forward on to the hard, snow-covered earth. His face met the ground with a soft thud but that did not – could not – render him unconscious.

He tried to shout out his fear but it was too late. His facial muscles no longer responded, only the mental processes allowed him the feelings of sadness and helplessness. They were for her... and was it for their son?

He could see along his arm, to his fur-mittened hand lying outstretched on the frozen earth. He could see a faint glow dancing around it, shimmering, blue-white and opalescent, encompassing first his hand, then slowly advancing up his arm before engulfing his whole body. He tingled all over before he felt the first tremor, then another and another before blackness, calmness and peace descended.

Out of the darkness he saw the brilliant pinpoint of white light at the end of a long tunnel. He heard the voices calling him, pulling him through. He was no longer afraid for himself. Only for Ganhumara and a Destiny that now could never be. He was cheating mortal death. He no longer feared it, instead 'fear' had been replaced by 'exhilaration': that he was passing from one phase of existence to another. He now understood what he had never understood before: it was the mind that operated the gateway, that it was in the quantum thought processes of the mind and of the soul which allowed instant access to any place and at any time.

Was this not what all artists, all creative people, instinctively knew? Was this not where imagination had its cradle? That all times past and all times future were 'now' this 'one time'?

Events only had meaning and existence within the minds of other men: an event ceases – and never had existence – when the last man forgets it. What men would actually see, would remember of him, had already been written. It was only an image of him, a part of the substance that already formed the whole universe. That image was already imprinted on the endless loops and coils and twists of time, locked into sets of actions and reactions which would never vary until the universe itself collapsed back into the point source, the quantum event, that had been its beginning and which would one day, billions of years in the future, would be its end.

Except for the soul. His soul was about to make its transition. It was moving from darkness into light. That real part – the essential part that had made himself, himself – was now escaping the rigour of time, was stepping outside its physical confines.

Like a deep tugging of the heart his soul started leeching away, wanting to fuse, to be part of someone else so that the concept of 'self' became irrelevant. Ganhumara had somehow always known it of him.

Without looking she had found and held it and loved it, without asking or wanting anything in return. Only she was no longer called Ganhumara. She was someone else. But who? Where had she come from? Where was she now? Would he still be a part of her? Would he ever again see her and feel her and need her?

Now he was going out to be reassembled among the beauty of the stars where he would be part of them and they would be part of him. He could see the Creator's face with its infinite kindness and grief and understanding and love. In his mind he could see Ganhumara's face too, feel her heartbeat along with that of the universe. He tried to reach out and touch her. The process had begun. He was returning to where he had come from.

∞

Dr Orum looked up from writing his patient notes. There was a newcomer. It was consultant neurobiologist, Dr Richard King. 'Hi, Richard!' said Orum. 'This Dr Ortez should be arriving anytime soon. Do you know about it? Orum's question did not get an immediate response as King continued to take his coat off. Orum continued: 'The medical director's clearly excited about this guy. He's been behaving like a cat on a hot tin roof for the past few days. Been down here several times. So I reckon Dr Ortez is important.'

King was a tall, balding Scotsman in his early forties, with a formidable reputation in his medical specialism. 'How could I know anything? I've only just got here,' he replied at last.

"Yeah, sorry,' replied Orum.

King continued putting on the spare white lab coat hanging up behind the door.

Orum continued. 'All these 'new' people coming and going.' He leaned back in his chair. 'There's going to be a cast of thousands here soon?'

Did they tell you I was coming?'

'No. Not a sodding word.'

'Okay. Sorry about that,' said King. 'Breakdown in hospital communication.' He paused as he finished buttoning up his lab coat: 'What about this guy from the States?'

'The guy we're expecting? The one from Los Alamos?' queried Orum.

'Yeh. The new guy.'

'He's the top man. Name's Juan Ortez. Travels around the world, visiting universities and hospitals like ours. There's evidently twelve such facilities like ours worldwide.' Orum pulled a face.

'I didn't quite buy it though. Somehow he's connected with the other American, Coleman is it and the creepy British guy, Sir Freddie somebody or other. They're all coming here. Should be arriving any minute.'

King sat down next to Orum.

'God, I could just do with a cigarette,' said Orum. 'You haven't got any, huh?' He looked hopefully at King.

'I don't smoke.'

' And neither does he,' Orum replied nodding in the direction of his silent medical junior writing up his ward notes.

'God! As neither of you can help I'll bugger off and buy myself a packet from the kiosk downstairs.'

As he rose to leave King asked. 'So who's the new guy? What does he do, and why do you think he's coming here?'

'I told you, he's their Mr Big. The director said he'd stressed in an email he'd received the importance of prepping patients properly, to get the best out of their imaging system. He's going to show us how to do it.'

With those words, Orum disappeared through the open door saying loudly as he left, 'Whatever they're gonna do I'm off to get my nicotine fix first.'

He'd only been gone a few moments when the medical director bustled into the room. Behind him was a man of small stature and dark, slick, brushed back hair with a suntanned complexion. He was carrying a large, leather bag festooned with airline flight tags. It was heavy judging from the effort it took to place it gently on to one of the room's still vacant chairs.

'I was expecting our other American guest to be here by now along with Sir Freddie,' said the medical director in a slightly irritated tone. 'No matter. Let me introduce Dr Ortez, whom we've all been expecting but whose arrived a bit earlier than our other guests.'

As he finished speaking a red faced Dr Hugh Orum reappeared. 'Sorry I was gone,' he said to no one in particular. 'Had to make a call of nature.'

Dr. King along with Orum's assistant were sitting in two of the chairs opposite the still-blank computer screen. As King got up to shake Ortez's hand another junior doctor entered along with the hospital's consultant psychologist, Dr Tim Lord.

King followed by Orum were the first to accept Ortez's proffered handshake. 'Hello,' Dr Ortez said in American English, with a hint of Spanish accent. 'Nice to meet you all.

Hope I can be of some help. I was hoping to meet a Dexter Coleman too and I see he's not here. But I'm early. Our embassy has contacted him. Before I do anything I've been instructed to wait his arrival.'

The medical director said: 'We've been asked by the American government, the White House no less, to extend Dr Ortez every courtesy and aid while he's here. Our own prime minister's office has cleared all this.' He paused momentarily. But he did of course, stress that anything we do must not threaten the health or safety of our patients.'

'I'm sure Dr Ortez's visit is going to be informative for us all. It's not often we get someone as senior visiting us, from the US's most prestigious research lab.' He paused for breath before continuing. 'I'm sure I speak for all of us when I say it's an honour to have him here.'

The medical director turned to Dr Ortez. 'Would you like a cup of tea or coffee whilst we wait?' He'd had no sooner said it when Sir Freddie and Dexter Coleman arrived together.

Sir Freddie said brusquely. 'Something cropped up. Had to meet the PM. Dexter here had a long telephone call with the president.' He did not elaborate further.

'Good to see you back, Sir Freddie, the medical director said adding, 'Did you inform our patient's next of kin?'

'It's in hand,' replied Sir Freddie in a manner that brooked no further discussion. That the two newcomers had been speaking to either the president or the PM was not lost on those present.

'Now, if we could just go through to the ward with Dr Ortez?' said Sir Freddie. It was not a question. Without waiting for a reply he was already heading toward the imaging ward. Orum automatically began following suit but was stopped by Sir Freddie. 'You can take it from me that Dr Ortez knows what he's doing,' he said. 'But spectators might make him nervous. Perhaps it might be better if you and your team remain at the computer end of things monitoring the results?'

'Good thought,' answered Orum hiding his irritation.

Orum remained standing so that he could see across the corridor into the small ward containing his bedridden patients. He could see too Dr Ortez talking to the American. The conversation continued for a few minutes with Sir Freddie for once remaining an idle bystander. He saw Dr Ortez place his bag on a convenient chair close to the bed of Dr. Jonathan Anderson.

He saw Dr Ortez open his large leather bag then take from it a box containing various items of medical paraphanalia, and a packet of cotton wool. He broke off a small wodge of the cotton wool and then took the limp arm of Jonathan Anderson. He rolled up the pyjama sleeve covering his left arm. Quickly removing a large bottle of liquid from his medical box he swabbed the arm. Then he saw Dr Ortez withdraw what Orum presumed to be a special and already pre loaded disposable hypodermic syringe. Ortez held it against Anderson's arm for a second or two before laying the empty syringe onto a nearby table. He murmured a few words to the American before the three of them left the ward to return to the Imaging Room.

In answer to an unasked question hanging in the air Dr Ortez said. 'It's something we use to prepare patients for imaging.'

'Yes,' said Orum as if in agreement.

'It's something we've developed in the 'States. We find it can gives us very good results.'

Coleman and Nightingale had been slightly behind Ortez, talking quietly to each other. They rejoined everyone as Ortez added. 'It needs about twenty minutes or so for it to take effect. Then we'll give him the booster shot.' He paused. 'We have decided to use Dr Anderson first but for no particular reason.'

'We supplied your hospital with only one of our special SQUID helmets. Perhaps one of you or a nurse could get it for us so that we can fit it on to Dr Anderson's head?'

Dr Ortez surveyed each of the faces staring at him. 'Do not worry, this technique is virtually non-invasive, as you saw. No one will come to any harm whatsoever. I guarantee it. We often find it has beneficial side effects.'

It took five minutes before a nurse carrying the SQUID helmet appeared. Dr Ortez took it from her. They watched as he and the nurse went back across to the ward where he expertly fitted it around Dr Jonathan Anderson's head, taking care not to disturb the various other wires and tubes, attached to the patient's body.

There were a large number of electrical leads that both he and the nurse re connected to the helmet. The whole process took over twenty minutes. Finally, they were finished. The nurse left and Dr Ortez checked his watch. Then he felt Anderson's pulse. He seemed satisfied before taking another hypodermic syringe similar to the first one, from his medical box. Again he swabbed the patent's arm before injecting him with the second liquid.

Ortez straitened up from his task then placed the second syringe alongside the first. He closed his medical box, closed his bag and then rejoined his audience for the second time.

He commented to them, 'I have to warn you that the results could take some time.'

'Don't worry about me, Doc, it's why I'm here,' said Coleman. 'But could you not have done the woman first?'

'No one gave me any instructions who to do first and in what order,' Ortez calmly replied. 'It's too late now since I've prepped the man.'

'Any idea how long it usually takes?' Urged Coleman.

'About three hours is normal.'

'You'd then move on to the woman?'

'Yes. But it would depend on the results we get from the man of course. But that's basically how we would do it,' answered Ortez.

Ortez looked at his watch. We could have them all done by say …' he looked at his watch again '…by, I would say, two o'clock tomorrow morning.'

'And I'll be staying for as long as it takes too,' added Sir Freddie.

Orum found himself looking at the still blank computer screen.

∞

Anderson fought hard to form the words as he lay in bed. In his mind he could hear those around him yet, for reasons he could not understand, he could not answer them, the words would not come from his lips. He continued to struggle.

'My name is Jonathan Anderson … Dr Jonathan … Anderson … I am a scientist … a nuclear scientist … They know me … of course, they know me … at Mary Magdalena Church, Ripley, Surrey, where I preach … My name is Jonathan … Doctor Jon … My name is Jon …'

He accepted that the ability to speak had slipped from him. It was no longer any use to try. The words would not come. He doubted they would ever come again. Then he heard the other voices, the strange voices from across a gulf of time, from across a vast void of nothingness beyond his ability to comprehend.

A serene calmness permeated the deeper recesses of his soul. He felt like a small boat on a river where he had been rowing against the current, for so long it felt like all eternity. Now, he was tired, exhausted, so exhausted he could row no more. The current could carry him to wherever it chose … to a distant shore, a distant land. His struggle was over. He was at peace.

The first voice he had heard was still there, still far off. It was unusual in a way he did not understand, even though he instinctively knew it was close to his own native tongue. It had an accent he had not heard before; foreign, yet somehow familiar. Not Saxon. Not Celtic, either. What was it? He knew. Yet, at the same time, he did not.

∞

Images flashed on the computer screen, while Orum and his small audience looked excitedly on and, more often than not, marvelled at the clarity of what they were witnessing. From time to time some would cross the corridor to where the patient lay, to ascertain whether there had been any change of the patient's condition. Having satisfied themselves that everything was the same, they would return.

∞

Yet mysteriously, Anderson could still see and hear the voices of that other place and that other time. Somehow, he was both in this place and in that other place, too. How was that possible?

He heard another deeper and tenser voice from that other time.

'Has he found them yet? Do you think he's found them?' There was an urgency, a pleading.

A second voice replied: 'We don't know. It's very experimental. We don't know the damage we might cause. It could be irreversible.'

'Goddamn it!' said the deeper voice, even more agitated. 'There are several million US lives at stake. Do what you have to. But do it now!'

'It takes time,' said the first voice, firmly. 'He has to have the time he needs.'

'I don't have time. My countrymen are fresh out of time. They can't wait,' retorted the deep voice.

The man once called Dr Jonathan Anderson could see them all now. The vision of his dream of a few moments ago had cleared. The steam, the water, the grey and granite walls were gone. Slowly, he was rising up, off the bed where he had lain, looking down on the scene below, at men dressed in strange white clothing. He wanted to shout: 'I'm here! Look up! This is where I am!' but the words still would not come. The same mysterious force that had already paralysed him was not freeing him but trying to propel him towards a distant and familiar past. He was slipping towards it but he knew that this time it could not hold him.

Then it was quiet. The place of wizards was gone. He was in that other time, that other so different place. He welcomed being back where he belonged. Somehow and somewhere he had crossed a border dividing the two. He was in that familiar other time, that other place of ... of ...? Was it 1,500 years in the past? an inner voice asked. The time of Ar-tur? After the Romans had gone? When he had fought the Saxons and had won ... and lost ... both at the same time?

He felt the familiar, comforting warmth of her body. Together they lay in the stillness of the bathhouse, the place he had always known as the temple to Aquae Sulis. In his mind's eye, he could see it from above, even though he knew it was impossible.

'I have to give you these,' she said. 'They will protect you. They will protect us, and the world yet to come.'

She withdrew from his embrace, placing his arms to one side. From beneath her white Roman stolla she withdrew the metal objects and placed them in his hand. They felt warm, alive. 'You must throw them into the waters, as the Romans once did.'

He looked at her, then at the hot pool at their feet, then at the objects in his hand. He saw the bottom of the pool, he saw the glinting of the many coins and jewellery item cast in by all those who had come before. He looked at the objects, then at the amulets he held in his hand, for the last time. The inscriptions were identical on both amulets:

Property of the United States Government.
If found return to Los Alamos National Laboratory.
Shipping address: PO Box 1633, NM 87545 USA.

Serial No: 175890/1603/11.

Possession is an offence without proper authorization.

The warlord drew his arm back. He turned to her with a puzzled look on his grizzled face. 'I feel we are being watched by an unseen presence.'

'It has been the same for me, ever since leaving Ynys Witrin.'

The warlord did not reply. With a shout, he drew his arm back, then brought it swiftly forward, releasing his hold on the amulets. He and his queen made their wish.

The amulets curved high into the air to be swallowed by the darkness, before descending and splashing, finally, into the steaming waters of the temple of Aquae Sulis.

Then they turned and began walking out into the cold, crisp air of that winter's evening. But it was becoming darker. He was sliding back, as he knew he must. Back towards where he had come from. He held out his arms for Ganhumara. She tried to hold him. But he felt himself slipping away.

∞

There was silence from the team of doctors and from both Coleman and Sir Freddie as they had watched the scenes unfolding before them on the computer screen. One of the doctors murmured almost immediately as he watched: 'Mmm. The Roman baths look much the same today. Aquae Sulis? Aquae Sulis? He repeated. 'Jesus! That really is amazing.' It was followed by silence from everyone. Then the doctor, in an effort to fill the vacuum he'd created added. 'How strange. My wife and I were there for our summer holidays.'

Dexter Coleman heard him, but left it for a few moments before picking up on it.

The door to the Imaging Room opened and Dr Orum reappeared after going over to the ward to check his patient. 'I don't know what you've got over here, but there's no physical change.'

Coleman appeared not to have heard, for he was more engrossed in the doctor's comment about his holiday.

'What do you mean, you were there?' Coleman asked at last.

The doctor was surprised by his question. 'Well, it's obvious to anyone who's been there. For a start, it's the only place in the British Isles with hot, underground water. There's nowhere else. So it's simply got to be the Royal and Ancient City of Bath.'

Coleman looked perplexed.

'Fascinating museum,' the doctor continued. 'They've been excavating those Roman baths for the last three hundred years. They've done an amazing restoration job. A wonderful experience going round it I can tell you.'

Sir Freddie had caught the hint of excitement in the American's voice. Sir Freddie knew Bath well for he'd once been its Member of Parliament. 'It's one of our oldest cities.' he explained. ' 2000 years old. Population now about 100,000. Beautiful place. It's about a hundred miles due west of London, in the county of Somerset. Very old and beautiful city. The Royal Crescent with its line of beautiful, Nash designed houses. Jane Austen used the city many times in her books.' Sir Freddie looked inquiringly at Coleman. 'Have you read any of Austen? "Pride and Prejudice", "Northanger Abbey"?

'A long time ago', Coleman replied vaguely.

Sir Freddie knew he was lying. It was unimportant. He ploughed on regardless. 'A UNESCO World Heritage site I now believe. Established by the Roman's 1500 years ago who first called it Sulis Minervae although it was known by the native's long before. Is there something special about it for you?' Sir Freddie enquired.

'Could be. Could be,' Coleman replied. At the same time he tried hard to recover his usually impassive face as he tried hard to hide his belief that he might have struck the mother load.

'What's the best way of getting there and how quickly can it be done,' he tried to ask casually.

'Driving straight down our M4 motorway is the way I would do it,' Sir Freddie replied before asking. 'Could you drive there? Or could you be driven?' Sir Freddie then added, 'Failing that it would be the rail connection. It's a very good service and very fast from Paddington, one of London's major rail stations.'

A sudden quiet fell on the room. Sir Freddie was the first to speak again. 'King Arthur is supposed to have fought his first major battle there at Mons Badonicus where he defeated the Saxons so soundly the peace lasted a decade or more. That was until they came back and so legend has it, defeated Arthur at somewhere called "Camlan" wherever that was…'

But Coleman wasn't really listening. Suddenly there were things for him to do and places for him to go. He looked vacantly around at the faces staring back at him. Coleman knew what he had to do and he knew he had to do it quickly.

CHAPTER 25

It was the place Ganhumara had come to dread. It was the place she knew as Camlan. It was a place in the heartland of Saxon territory not far from the defensive fort line built by the Romans along the southern coast to keep Saxon invaders at bay. It was where her warlord had finally caught up with Saxon King, Aelle. It was where her warlord had chased him after a series of running fights beginning near the site of his former victory many years ago, where the seven hills surrounded Aquae Sulis; on the hill they called Mount Badon.

She had always known what was to come, what would make the final battle site a lost place of legend, where Ar tur would receive his grievous wound. He would survive, but only long enough to be brought back to Ynys Witrin, to the shore of the inland sea.

She felt again the cold hand settle on her heart. Love would not be enough. Not this time. She could not change fate, despite already knowing of what was yet to come. She could only hope against hope the amulets she had given him, which he had tossed into the hot spring, would protect him, might keep him safe, locked within the endless loops of time. Even if she did not see him again, it would be enough to know he was safe.

She saw the figures in the snow moving away from the once-heaving mass of bodies, the blood and gore splashed in the white snow of the battlefield. She saw the scene cloaked in silence, as if watching a silent movie from an infinite distance ... like a dream, a nightmare. And what was a silent movie? She was propelled onwards by the same irresistible force, moved to the inland sea where she saw the same four figures moving wearily towards her. Then she saw the black barge with its white sail moving across the inland sea, before making its last journey back to the island of St Brigid. She saw the cloaked figure of herself, as she had once been, and of some of her helpers from her order of healing.

Now, the figure in the barge and herself had become one. She was close to him, looking down into his eyes as he looked up into hers. But the light was fading. Something was leaving her, their souls were merging as they became one. Yet he was safe, despite the flickering of his life-force.

As he lay in the barge he beckoned one of his men over ... asked him to take his sword ... the one Myrradin had made for him according to instructions from Ar-tur. Together they had forged it, but its secrets could never be revealed in this time. He beckoned one of his last surviving soldiers over and told him what must be done.

Then the barge pulled away from the snow-covered shoreline. She was now no longer on the barge. She was alone in the white silence where she could hear voices from a long way off. There were things for her to do. And there were people waiting for her to do them.

∞

For Dexter Coleman, the 110-mile drive south-west along the British M4 motorway had proved straightforward. It had been three and a half hours since he'd left London, heading for the UNESCO World Heritage-denominated site of the city of Bath. His car's satellite navigation system had done the hard work and now his journey was almost complete. Ahead, through the car's windscreen, he could see his hotel, the Bath Hilton, where the embassy had arranged he could stay overnight should the need arise. That would depend on how much stamina he would need in order to get done what he planned and then drive back to London. The most important aspect was meeting the president's deadline for DEEP EARTH. He had plenty of time according to his own reckoning, but everything was a moveable feast. Nothing was certain. At the moment it all depended on how the next few hours panned out. His fingers were metaphorically crossed. Already his day had proved long, and it was only midway through the afternoon.

His day had begun when he'd arrived back at the embassy in the small hours of that morning from the Atkinson Morley care of Sir Freddie Stirling. They'd talked on the way back while his own embassy-provided car had been driven back empty. Before grabbing a couple of hours' sleep he'd lost no time in getting the embassy's night staff to make the necessary preparations for his expected journey to the city of Bath later on during that same day's afternoon. Despite embassy opposition he'd decided to ask them to provide him a rental car so he would drive himself. He believed he would be coming close to that 'make or break' time he'd once been familiar with on operational duty in the past. If it was to end in failure he didn't want a witness.

The Roman baths were in the centre of the city and only a short walk from the hotel. During his drive he'd taken the phone message from embassy staff:

'Park the car at the Hilton! ... driving anywhere in the city is a nightmare ... it's a one-way system.'

Coleman hoped his mission would be over quickly. Because of the urgency his trip carried, he might be there a matter of a few short hours, before travelling back to London. It was a hope, maybe a false hope, but long experience had taught him that 'things' were hardly ever straightforward. That old saying rang in the back of his mind: 'if things can go wrong then surely they will'. But White House Chief of Staff Cliff Wurzburg had underlined the urgency during an earlier discussion that day, almost as soon as he had woken up. 'The president is under a lot of pressure from the military to launch alternatives to DEEP EARTH,' he'd said. 'So you've got to be quick on this. He might not be able to resist the pressure for much longer.'

In the car, during the drive back from the Atkinson Morley, Sir Freddie had proved a mine of information both about the city of Bath itself and its Roman baths in particular. He'd told him how they had been discovered as recently as 1755, since when they'd remained extremely popular with the public. He'd sought Sir Freddie's advice on the quickest and easiest method of getting there, learning it was a toss-up between using the train or driving.

After Sir Freddie had dropped him off, he'd realised it was too late to make his briefing call to Wurzburg, even after allowing for the five-hour difference between London and Washington. He had no option but to leave it until later that same morning, but Wurzburg had instead gotten to him.

Wurzburg had looked grim but had brightened considerably when he'd revealed his latest news concerning Bath and its museum.

'Great! Great! That's really great news, and we need some. The president is getting edgy thinking about DEEP EARTH and what he might have to do if you don't come through. It's a big task, I don't mind telling ya; one that he's not keen on doing if he doesn't have to. I don't need to add more, Dexter.'

Dexter was a pro and knew what he had to do. As he'd left London he'd realised he'd left the embassy to make all the necessary arrangements. His last task had been to speak with the ambassador himself where, without revealing too much, he'd impressed on him the importance of his trip. 'Matter of life or death,' he'd said. 'I don't care where the curator is. I gotta see him. Make sure either they, or their assistant, are there.'

The ambassador had taken it all as if it was an everyday occurrence. He'd merely nodded as if he'd heard it all before, saying it would be done 'come Hell or high water'. With that, Dexter Coleman had been on his way out of London.

Coleman had thought long and hard about the chauffeured embassy limo, complete with the security detail he was entitled to. But entering Bath with an entourage was not the best way of avoiding attention. It wasn't a question of being secretive, it was more of avoiding broadcasting his presence. All it would take was some inquisitive newspaper journo knowing who he was, and then he could be in a sea of trouble, with questions asked about why America's intelligence chief was there. In no time, it could end up on the front page of the New York Times, getting him no thanks from the White House. So, he'd erred on the side of caution.

He'd told the embassy to make his appointment with the curator for 'when I get there'. He'd checked his watch and asked, 'When do you think that might be?'

'Best leave four hours' travel time,' they'd said.

'So, four o'clock would be about right,' he'd murmured back.

Arriving at the Hilton he'd checked his watch at the reception desk. It was 3.45 p.m. He asked the young female receptionist how long it would take to walk to the museum. 'Fifteen minutes,' had come the courteous response. He'd then asked her for directions.

The air was cold as he walked. He noted that traffic seemed busy and was almost choked to a standstill by the frequent traffic lights. A light rain began falling, making him glad he was wearing his long, charcoal-grey overcoat and red scarf. In less than fifteen minutes he was at the entrance to the Pump Room that, from university days, he remembered had been much-mentioned in Jane Austen's many books based on the city. In particular, he still remembered her seminal novel Northanger Abbey, and recalled that it had been the Pump

Room where the heroine, Catherine Morland, had met the hero, Henry Tilney.

Despite being preoccupied with matters in hand, Coleman could not help but be touched by the architectural beauty of Bath as he walked, and which gave him a brief insight into why it had been designated a UN World Heritage site.

Coleman's coat was slightly damp from the lightly falling rain as he walked in through the Pump Room entrance. A single, uniformed commissionaire asked him who he was meeting. With pleasant surprise, he noted that the man had his name written in a large, official-looking and ornate book where many other names were also entered. He was asked to wait a few moments while the commissionaire telephoned an anonymous person. After a few seconds he replaced his receiver and then murmured politely: 'Someone will be down to collect you.'

A lady in her early fifties appeared, and asked him to follow her. The journey took them past ongoing building work, through a large kitchen, with staff midway through preparing meals for later that day, then through the actual Pump Room itself: a large, high-ceilinged Georgian-style room of great elegance with a string quartet playing unobtrusively at one end. The room's overall décor was amplified by the whiteness of the walls and glowing chandeliers, offset by a period black-and-white chequered floor that provided space for at least a hundred tables, filled mostly with people taking afternoon tea.

Finally, they emerged into what appeared to be the building's original entrance and ticketing hall. His guide suddenly disappeared through an anonymous side door off the ticketing hall and leading up several flights of narrow and winding stairs. Finally a landing opened out on either side of the staircase to what Coleman guessed must be their destination on the top floor. There were two offices, one on either side of the landing and both with no doors. From where he was standing both looked tiny, crowded with paper, computer screens, books, files, ancient desks and office furniture, all arranged on top of each other. There was barely enough room to squeeze through. A brief thought concerning fire regulations went quickly through Coleman's mind.

The smaller of the two rooms had windows that looked out and down on to part of one of the hot baths. Seated at the furthest desk was a man who Coleman assumed could only have been the curator. He didn't get up to greet his guest but remained seated, still mostly absorbed in what he had been doing before Coleman arrived. Eventually he looked up at Coleman.

'Hello,' he said finally and quietly. 'We've got to help you, I understand. We've had a few calls from your embassy. The ambassador himself, I've been told. I don't think we've had that before. Now, tell me, what can we do?'

'My name's Guy Kendrick,' the man added in a low, softly spoken voice. 'Been here for the last thirty years. Not much I don't know about this place, or our exhibits.'

'I guess my secretary, or maybe the ambassador, told you why I was coming?' said Coleman.

'Yes, he did. Said you'd lost something. Thought that, somehow, we might have it, though don't see how. He said you'd explain.'

'That's right. I can't tell you how it might have got here, or what it is. All I can say is what it looks like. There are two of them, and they're small, dull, grey metallic-looking … possibly joined together by a chain. There's an inscription on both of them; words to the effect of "Los Alamos". Should have maybe "Property of the US Government", as well.'

'Would that be in English?' the curator asked.

'Yes, in modern English,' Coleman replied, understanding immediately its possible incongruity.

'Seems unlikely,' said the curator unhelpfully. 'You see, everything from the period would have been written in Latin, not English. English, as a language, came hundreds of years later, well after the time of the Roman baths.' There was a puzzled look on his face.

Coleman understood his logic. He tried another tack. 'For the sake of argument, let's assume that we know the discs are here, somewhere. They were definitely last seen at your Roman baths. How they got here is a long and incredible story, but just go with me on this.' Coleman checked the look on Kendrick's face. He took his silence as tacit acknowledgement.

'Say they'd been lost here. Thrown into the hot springs ...'

A spark of recognition appeared on Kendrick's face. 'That would have been in what we call the Sacred Spring. There are two pools, you see. That's the one where the Romans made their votive offerings to the gods.'

Coleman didn't want to listen to more than necessary. 'Would everything have been catalogued?' he patiently asked.

'Oh, yes. We have two periods. The first stems from 1879, when Major Charles Davis did a lot of excavating, but that was a relatively small bit of the actual baths. The next was in 1970, which was far more extensive and well-catalogued.'

'They went through the floor of the Georgian baths originally built on top of the old Roman ones. There was a difference of about eight feet in level.'

'Excellent, really excellent,' said Coleman, sensing that he might be on to something. 'These discs would have been beyond the technology of the 1970s. They would have been as incomprehensible as finding a television set in 1879. Even though the discs might have been recognised as being from the wrong era in your excavations, would you have kept them, or dumped them with the trash?'

'We were meticulous,' responded a seemingly affronted curator. 'Everything, and I mean everything, that was excavated would have been catalogued. If the excavators of the time had not known what an object was, they would have catalogued it simply as "unknown votive offering".'

'Is it possible to look through your catalogues?' asked Coleman, trying to hide his excitement.

'Absolutely. It's easy. Our catalogues are on this computer in front of me. And each and every item has been photographed.'

'Could you look through it for me, with me looking over your shoulder, as I know what I'm looking for?'

'No problem. How should I look?' Kendrick asked. 'I could enter "Los Alamos, votive", plus "disk", in the computer's search box. You never know, we might strike lucky. Beyond that, it could be a lengthy process.'

'Let's try. You never know. But suppose we do strike lucky, where would you store them?'

'Well, they could be on display either in the Roman baths themselves. But I rather think, given your description, they'll be stored in the room next door. You see those plastic containers?'

Coleman twisted round to look into the room behind him.

'Yup, I see them.'

'That's where they'll be.' He paused as he started tapping away into the computer. 'No, nothing.' He turned to look at Coleman, and then looked back at the screen. His brow furrowed. 'I think I've spelt Los Alamos incorrectly. I spelt it with a "z". Hang on …' He typed away again. A picture came up. 'Is that them?'

Coleman looked over the curator's shoulder. There was a photograph of both sides of the objects. One photograph clearly showed the inscription:

> *Property of the United States Government.*
> *If found return to Los Alamos National Laboratory.*
> *Shipping address: PO Box 1633, NM 87545 USA.*
> *Serial No: 175890/1603/11.*
> *Possession is an offence without proper authorization.*

'How curious,' said the curator. 'I don't think I've ever seen them before. No reason to, really. I wonder what the archaeologists made of them when they were first discovered?'

Coleman's first thought was to ask why the museum had not returned the discs to Los Alamos? There could have been a million reasons. But what did it matter? What mattered was that they were here now. His search was over. His hands were trembling and they hadn't done that for a long time. All thoughts of the day's fatigue had gone, to be replaced with thoughts of getting back to the White House in the shortest possible time. Maybe too he might get the New Mexico state police to organise a meeting with Pahlavi still being held in their cells. Maybe too with travelling back via Los Alamos with Doctor Ortez and heading straight for their nuclear imaging facility. It could greatly quicken obtaining the truth from Pahlavi as to whether he'd 'doctored' his results or not. It was a possibility.

∞

Compared with the excitement of the previous day and early morning, the Imaging Room at the Atkinson Morley hospital was quiet. A nurse had just finished making her rounds, checking the vital signs of the four patients still occupying the beds in the only ward. For the past twenty-four hours nothing had changed, as it had not changed since the patients' first arrival. They still lay asleep, but otherwise in perfect health.

Dr Jonathan Anderson lay in the first bed, his wife next to him and their son, Ben, in the next bed. Furthest from the door lay Ben's nurse, Dolores. There were no sounds, not even that of breathing.

Across the corridor only one doctor remained on duty. He was scribbling in a pad. The telephone rang. He picked it up and murmured in response to a question being asked on the other end. He replaced the receiver and then, within a few seconds, left the monitoring station to pursue an unknown errand.

For a few minutes, no one was there to hear the sound of a child muttering a nursery rhyme:

> Star light, Star bright,

it began, haltingly, for a few moments, then continued:

> First star I see tonight;
> I wish I may, I wish I might,
> Have the wish I wish tonight.

The voice halted for a second time.

The only other person dimly aware of it was Dr Jonathan Anderson. To him it was as if he were witnessing it from a long way off, from another place, far away. But the child's voice would not go away. It grew closer and closer.

He opened his eyes. There was a face, a small and beaming face. It took only a few seconds for him to realise it was Ben's. It took a few more seconds to realise his son was singing. There was no mistaking the look of happiness, the sound of profound joy in his voice. Dr Jonathan Anderson was surprised, for he had never witnessed it before. His son was smiling and singing, and smiling again, as he looked down on his father.

'Get up, Dad,' his son said. 'It's another day. Mum's asleep.'

Then another face came into view. It was that of Dolores. Slowly, everything was coming back. It had been like a dream where he had been trapped in a place far away.

He felt life begin to flood back into his limbs.

There was someone missing still: a face necessary to complete his return to reality.

CHAPTER 26

Everyone accidentally caught in the beam's "light" at Pendragon Point had been affected in different ways although each of those affected did not know this. Some were more affected than others. Katrina and Jonathan were the worst. Ben had been affected differently. His carer, Dolores hardly at all. Why this happened would take years of nuclear physics and medical research to explain. But the principal impact on Ben's parents had been to fuse together their appreciation of time into a continuous stream of 'one time'. It was as if taking off in an aircraft and looking down on the runway lights. Each light represented a moment in time. The higher the altitude the aircraft attained over the runway so each individual light seemed to fuse into the next until all lights could be viewed at the same time as if stationary.

The major question they would both be asked for years to come was whether there was any reality to any of these times. Certain indisputable facts seemed to indicate there was.

Katrina was acutely aware of happenings around her hospital bed but they seemed to her to be a long way off and she was remote from them, unable to influence their outcome. She was a shadow there, a ghost, a wraith.

The past was altogether different. That was where her life now was. It had colour and meaning. More, she was living that life as if she had always been there. Yet always in her mind was that feeling that it was somehow temporary. That at any moment she could and would one day be called back to her 'other' life.

Part of it was Jonathan. Katrina was aware she loved him in all times whatever his name or fate might be. Her love was unconditional and crossed seamlessly all time lines. Without hesitation she always accepted the way things were and that the Jonathan of the hospital and her Warlord were inexplicably connected. At that moment as she lay in her hospital bed her mind was no longer there. She was without limit, without boundary. And it gave her a unique knowledge of times present and times yet to come.

∞

Ganhumara and Tomsantos stared up and along the narrow bridge of land connecting the mainland to the settlement on the far side. The bridge was shear granite on the side facing them with a hundred foot drop from top to bottom to where it met the ocean. Access to it was a hundred yards away across beachside rocks followed by a strip of grass-covered field. A hundred yards away lay the island that had once been her home for many years. Apart from the land bridge there was no way on or off. It was why the Warlord's ancestors had originally chosen it as their settlement.

'The only way to get us out is by starvation,' the Warlord had once said to her. 'And that could take forever given the ocean's plentiful supply of fish.'

It had been a warm summer's day devoid of wind. As they had stood surveying the island the iridescent sea had lapped at their feet and swirled idly over rocks on the other side before caressing the island's cliff face. To their left they could see the stiff climb to reach the land bridge and then a further hike before they could gain access across it and into the settlement. The path to it and across was wide enough, but only just, to take a horse and its rider. On either side it dropped into the sea. Ganhumara remembered how it had been when she had watched people toiling endlessly up and down the path as they had transferred cargo either on to or off the corbita then lying a short distance offshore. It seemed like a lifetime ago.

Shielding their eyes from the bright sun they scanned the island's flat, terraced summit. They could see the stone and thatch buildings, the low stone walls, the wooden stockade for cattle and the chimneys that were part of a number of tin smelters. They could see too the large building where Ganhumara had once lived. It all seemed as it should. Except there was no sign of life. It looked deserted.

'Looks like no one's at home,' said Tomsantos breaking the silence in a halting Latin he was still being taught.

'No, they are gone,' said Ganhumara but in an English that although she was still learning, she spoke almost fluently. 'I can only assume it is the Saxons. When I left a raid was feared as being imminent. It was why I accompanied you from here to Ynis Witrin on the merchant's ship all that time ago.' She paused as she continued looking for any signs of life, then said: 'Maybe the Saxons came as we suspected they would?'

'Maybe,' replied Santos. 'But what do we do now?

Myrradin was standing directly behind them both. Behind him were four Sarmation cavalrymen that had formed their escort as they had journeyed from Ynis Witrin a week's ride to the East. It was Myrradin who answered Santos's question. 'They have fled,' he said in an old English ascent only Ganhumara understood. She translated it for Santos.

Myrradin added: 'I will stay here with our Sarmatian escort. You two go on ahead and see what there is. Check that it is safe. Myself and the Sarmatians will wait for you until the sun sets. Then we will either meet you here or, if it is safe, signal for us to join you?'

'Sounds good to me,' said Santos but this time in his native American English. 'We yet may find what we came here to find,' he added. Ganhumara again translated for him but this time for Myrradin. Then she quickly added in English: 'I will follow you.'

The two set off briskly with Santos in the lead. It took almost an hour before they were on the opposite side and able to look back to where Myrradin waited almost half a mile away from their lofty vantage point. As they climbed they had found nothing to suggest a battle had been fought. And on reaching the settlement everything was normal. No burnt out houses, no upturned pots, no bodies. There were no animals either.

'An orderly withdrawal,' I would say,' said Santos at one stage as they looked around.

Ganhumara then took Santos to the point where she and her Warlord had once seen AURORA fall from the sky. She pointed to where they had seen "the light" originate, where they had been standing when it had wobbled and illuminated them both, and then discussed with him where the Warlord and his men had moved the object they called "the escape module".

'When the tide had washed it almost onto land, two strong swimmers swam out to it and attached ropes. They used these to drag it along to a sheltered bay close by. It was there they hauled it high up on the sandy beach, to where it would safe from high tides and prying Saxon eyes.'

Finally Ganhumara and Tomsantos made their way back from where she had once seen "the light" to the building where she had lived with her Warlord.

Inside she was relieved to find it almost just as she had left it. The rugs were still on the floors and hanging on the walls. The copper ornaments still decorated the interior, the furniture was where she had placed it. Overall it looked just as it always had: as if she still lived there. The only difference was the silence of the inside matching the silence of the outside. Both simultaneously became aware they had to signal Myrradin. They agreed it was safe to do so. Tomsantos was the one to make the signal. He would leave Ganhumara to find the artefacts they had come for, that had been taken from AURORA when carrying the dead bodies out from its interior.

After Tomsantos had gone it took Ganhumara only moments to locate the large wooden box in which she had stored the artefacts and items of clothing taken from the bodies. She saw it and bent down to open its lid. She looked inside. It was as she had left them.

Tom Santos had been in recovery for many months after AURORA'S crash. Fortunately he had been in excellent health before it despite being on the verge of retirement as a US Marshal but who still saw occasional service with both the CIA and the US National Intelligence Directorate. This factor alone underlined the high esteem he was held in along with the high value placed on his unique skills and experience. That experience had been further honed whilst working alongside US Special Forces deployed in the Afghan foothills of Kandahar province. He had spent long periods working on his own, liaising with some of the 1000 local villages in the area, blending in with local tribes of Tajiks, Uzbechs and Baloch, speaking their language, wearing their dress as part of a group trying to hunt down murderous Taliban opposition to Government forces.

He had remained unmarried after his first wife had died from a brain haemorrhage twenty years before. Her death had left him with a then young daughter for whom he had done his best to care for. She was now 26 and he was proud of her and her accomplishments. Over the intervening period their relationship had continually strengthened to the point where, despite occasional sexual dalliances when he was 'away on business', there was little else he much cared for.

He saw his 'day job' as a US Marshall being the end result of a gradual winding down of what had been for him a long and exciting career. It was a job that had taken him back to his roots in New Mexico, to his interest in the history and legends of the Old West. It had been arranged on his behalf by a Federal department so that he could more easily take care of his daughter on a day to day basis. But when push came to shove he could be called upon by the relevant institutions of a grateful country ' when and if the chips were down.'

It had all stood him in good stead during the months following his accidental crash into the past. That his mind too had remained clear had helped during his period of recovery when he'd had to contend with and accept the strangeness of his circumstances. His military background in survival and orienteering techniques had played a part.

They had helped him accept he was in a very different world from the one he'd left. He'd pieced most of it together over a period of time: the almost caveman dress of those around him, their strange and unintelligible language, the lack of any day to day conveniences such as a telephone or electricity. Although he remembered nothing of the crash, save for the blinding searchlight suddenly illuminating his brain before blackness descended, he'd still remembered he was flying to the UK from Albuquerque at the time. He'd been to the UK a few times before, enough to know that the UK he was now in was nothing like what it should have been.

For a start no one in this UK had initially been able to understand or speak his own language. He'd had to be taught the language, just as he'd once been taught Pashtun in Afghanistan, by living with the people with one of them accepting the roll of his language teacher. In this land that teacher had become a remarkable woman he'd come to know as 'Ganhumara'.

Wherever he was all he knew for certain was that it was nothing like what or where it should be. The only thing that still connected him with his past was his stainless steel Rolex Oyster Perpetual Submariner watch. The good news was it still worked. The bad was it didn't tell him the year and there was no way of guessing it. The best guess he could make was that somehow he was in a period several hundreds of years before his own. The only check he could make might be if he could retrieve any of the equipment left in or already retrieved by his rescuers from AURORA. He'd been told whilst he was in Ynis Witrin that they'd secured the wreckage high on a beach, safe from prying eyes, and that most of what was inside had been safely stored. At the time he'd had no idea of where or how far away it all was let alone how to get there. Unlike Afghanistan he had no modern aids to help, no local support, and no means of communication with friendly forces. He was truly on his own.

It had taken him months to learn sufficient of the language spoken by those around him to find out. Then had come the issue of convincing them to take him back to the wreck. It had been Ganhumara who had supported him from the beginning, almost to the point where it seemed that she too had something to learn by going back. The result was why they now were where they were. Maybe too, he'd thought afterwards, recovery of the wreck's contents might reveal a way of getting back to his own time. He'd discussed it with her just as they had discussed it endlessly during their long journey west from Ynis Witrin. Everything he'd seen during that journey underlined what he'd already guessed: somehow he'd ended up a long way in the past.

He'd slowly come to the conclusion that there was something 'odd' about Ganhumara, something that seemed to him to make her 'not quite fit' with those around her in the same way he himself did not fit. It was nothing specific, more than anything it was her manner, the way she readily accepted what he told her of the world he was from. It was as if she had once already known it but for some reason had forgotten.

That left 'Myrradin', the other person who'd taken a close interest in him and who again did not seem to quite fit.

He'd accepted that even to his own people Myrradin was regarded with a mixture of awe and strangeness, someone naturally distant and remote.

Ganhumara had once explained that it had been his long training necessary to become a 'druid'. This had been followed by his subsequent role as the tribe's priest, administrator of justice, healer, wise man and historian who could recite by wrote what had been passed down to him about tribal history, the land, the country, its legends and its culture. She'd explained something of druid history and Roman attempts to exterminate them as a race. How, unlike her own religion, they'd believed in different gods for different things and in the belief of the transmigration of souls; that this life was not the only one. There would be others. And with it a person's soul would be passed on from one life to another. It might in part explain why he too seemed to know things that were not of his own time or place, but from a remote future.

Myrradin had waited patiently as he had said he would until the signal from either Tomsantos or Ganhumara that all was safe on the island settlement. He sat on the rocks close to the water's edge whilst the Sarmatian escort waited a short distance from him. There were keeping themselves to themselves as they always did and as he'd expected. They sat muttering quietly amongst themselves whilst keeping their eye on the horses and provisions. They'd lit a small fire and had brewed a hot drink they'd shared amongst themselves. They were laughing and joking the way warriors do when relaxing.

Myrradin was the first to see the figure high up on the island waving down to them. He turned to face the Sarmatians and saw they'd seen it too. Rising from his seated position he walked over to join them. Nothing was said. Quickly all five were leading their horses first over the rocks and then along the path leading up to and over the land bridge and then into the settlement. They saw Tomsantos standing by the side of a large stone and thatched building waiting for them to join him. As they approached he ducked inside through an already open door.

Ganhumara looked momentarily at him whilst she continued with her task of emptying and mentally cataloguing the contents of the large box.

'Hello, she said simply and in English.

Tomsantos smiled at her. He saw all the items she had taken from the box and spread around her on the floor. He instantly recognised most of it ranging from sundry equipment to clothing originally worn by the other AURORA passengers. Amongst it all he quickly noticed his Glock 22 pistol. He wondered if it would still work. There was no reason for it not to. If it did work then they would have little to fear from Saxon or any other attackers he thought. There were other items too including the aluminium briefcase once strapped to the wrist of the Energy Secretary's Secret Service escort. With unvoiced regret he was unable to see the special, pocketed belt containing the two computer discs he'd once gone to so much trouble to save. Maybe they were under some of the clothing he asked himself? He would search as soon as the others reached him.

It was two hours later and Ganhumara and Tomsantos were seated on the headland virtually in the same position Ganhumara had once occupied months ago with her Warlord. They were looking out to sea just as she had been doing that fateful night when they had seen AURORA fall from the sky. She had just finished describing to Tomsantos what they had seen and what had happened.

'So that is what we'll do,' Ganhumara said at last, as she began to summarise what they had been talking about. 'You must write a message on a piece of slate or parchment saying you are here and need help. We will wrap it in your stetson hat maybe with some other item they will know is from you. That way anyone who finds it will know for certain the message it contains must be from you.'

Tomsantos shielded his eyes from the setting sun. 'Now it's my turn not to understand,' he said. 'How do you know anyone will find it?'

'I know,' she said. 'I know. Trust me in this. For with the assistance of "the light" I have already seen it. I will show you where to hide it at Aquae Sulis. Then, if I have to, I will tell them where it is. Back in your own time. Back in the time when I am something called a history professor...'

Tomsantos was not normally a man who would have accepted anything that he did not understand. But he was no longer living in normal times. He'd quickly reasoned that however unlikely what she was saying was, he was fresh out of anything better.

They continued to watch the sun as it began to slip below the far off horizon.

Ganhumara heard a bird began to sing high overhead. From out of nowhere she knew it to be called a "skylark". She listened *intently. Tomsantos motioned as if to speak. She brought a finger to her lips tacitly requesting his continued silence. As she listened she heard and saw the music well up in her mind with an elegance and simplicity she knew was not of her time.*

It was of another time, the one slowly seeping into her brain. Was that a violin she heard? What was a violin? Was the music by 'Ralph Vaughan-Williams'.

Who was he?

Could it be called "Lark Ascending"?

She blanched inwardly, frightened. It was the work of "The Light", for she had no clear idea what that word 'music' meant, or how a skylark related to it. But instinctively, she knew it was related to "that other place" and "that other time."

The music stopped and with it she turned to look towards the setting sun. She and Tomsantos saw the momentary flash as the sky closest to the sun turned from it's blood-red to a pale green, and then back to blood red again.

Was it a signal that her God was coming?

Or was it what Science would call in that "other time" "Rayleigh scattering" after the "1904 Nobel physics laureate Lord Rayleigh?" He'd explained it as "the scattering of different wavelengths of light by the Earth's atmosphere."

Ganhumara knew that she knew this from her days at Cambridge University. Although she was no scientist she knew its explanation. Yet for her the sight before them was in no way diminished by the explanation. She admired it for its stunning beauty beyond the ability of mere words to adequately explain.

CHAPTER 27

16.02 March 1st   Reuters News Flash

The US Department of Energy this morning announced the detonation of a 2-Megatonne nuclear device at their Hanford proving ground. Because the test was deep underground no radioactive contamination reached the atmosphere.

This first US test for many years, was conducted outside the current Comprehensive Nuclear Test Ban Treaty (CNTBT) adopted by the United Nations General Assembly, 10th September 1996, and signed by the United States but not ratified by it.

No further details were issued.

To be continued….

THANK YOU FOR READING THIS NOVEL

IF YOU ENJOYED IT PLEASE TELL YOUR FRIENDS AND WE WOULD MUCH APPRECIATE A "REVIEW" WRITTEN FOR THE AMAZON WEB SITE OR FOR WHICHEVER WEB SITE YOU BOUGHT IT FROM.

THIS IS HOW WE INDEPENDENT AUTHORS AND OUR BOOKS GET KNOWN.
Read the next book in the planned trilogy:
Book II: DEEP EARTH

The final book in the trilogy will be BRIGHTSTAR to be published in 2016.

Lightning Source UK Ltd.
Milton Keynes UK
UKOW04f1623170315

248045UK00001B/1/P